Building a Culture of Respect

Managing Bullying at Work

EDITED BY

NOREEN TEHRANI

London and New York

First published 2001 by Taylor & Francis
11 New Fetter Lane, London EC4P 4EE

Simultaneously published in the USA and Canada
by Taylor & Francis Inc
29 West 35th Street, New York, NY 10001

Taylor & Francis is an imprint of the Taylor & Francis Group

Typeset in Times by
The Running Head Limited, Cambridge
Printed and bound in Great Britain by
MPG Books Ltd, Bodmin

British Library Cataloguing in Publication Data
A catalogue record for this book is available from the British Library

Library of Congress Cataloging in Publication Data
A catalogue record for this book has been requested

ISBN 0–415–246–474 (hbk)
ISBN 0–415–246–482 (pbk)

Contents

List of contributors vii
Foreword by Lord Monkswell xi
Preface xiii
Acknowledgements xvii

Part 1 Origins and problems 1

1 **Origins of bullying: theoretical frameworks for explaining workplace bullying** 3
Helge Hoel and Cary L. Cooper

2 **Organisational responses to workplace bullying** 21
Neil Crawford

3 **Trauma, duress and stress** 33
Michael J. Scott and Stephen G. Stradling

4 **Victim to survivor** 43
Noreen Tehrani

Part 2 Size of the problem 59

5 **Social psychology of bullying in the workplace** 61
Claire Lawrence

6 **Monitoring bullying in the workplace** 77
Diane Beale

Part 3 Imperative to act 95

7 **Dignity at work: the legal framework** 97
Patricia Leighton

Part 4 The way forward 115

8 **A proactive approach** 117
Vivienne Walker

9 **A total quality approach to building a culture of respect** 135
Noreen Tehrani

10 **Don't suffer in silence: building an effective response to bullying at work** 155
Steve Rains

11 **Issues for counsellors and supporters** 165
Brigid Proctor and Noreen Tehrani

12 **Issues for general practitioners** 185
Chris Manning

13 **Bullying is a trade union issue** 201
Tom Mellish

14 **Building a culture of respect: issues for future consideration** 221
Noreen Tehrani

Index 225

Contributors

Diane Beal has researched work-related violence, in a variety of professions, since 1989, at the University of Nottingham. She has published widely in the area and has written national guidance. She has completed a PhD concerning the use of incident reporting systems and other methods of obtaining information about aggression and violence within organisations.

Cary L. Cooper is BUPA Professor of Organisational Psychology and Health at the Manchester School of Management, University of Manchester Institute of Science and Technology (UMIST). He is the author of numerous books and scholarly articles on issues such as occupational stress, women at work and industrial psychology. He is currently President of the British Academy of Management.

Neil Crawford worked as a psychotherapist acting as a principal consultant both at the Tavistock Clinic and the Institute of Directors specialising in human relations at work. He is co-author of *Bullying at Work* (Virago, 1992) and *Workplace Bullying in Japan* (NHK, 1998) and has written a range of papers on organisational consultation.

Helge Hoel is a research associate at the Manchester School of Management, University of Manchester Institute of Science and Technology (UMIST) where, together with Professor Cary Cooper, he carried out the first comprehensive nation-wide survey of workplace bullying. He is an active member of a network of researchers who have pioneered research into workplace bullying in Britain and have published several articles and book chapters concerned with the issue.

Claire Lawrence is a lecturer in psychology at the School of Psychology, University of Nottingham. She has researched in the area of work-related violence since 1990 and has collaborated in the design and delivery of training packages to reduce violence in the retail sector, the licensed trade, the prison service, public transport and the health service. She has also been a key partner in a EC-funded network examining violence and bullying in schools across Europe. Her research interests include the perception and understanding of violent behaviour, norms for aggressive behaviour and environmental influences on aggression.

Patricia Leighton is Professor and Head of the School of Law at the Manchester Metropolitan University. She is an employment lawyer specialising in health and safety at work, in particular the law relating to stress, violence and harassment at work.

Chris Manning worked in the NHS for 21 years, 15 of them as a GP. He has had major depression for the last 14 years. He is founder of PriMHE – a unique multi-funded partnership for mental health promotion in primary care.

Tom Mellish is Health and Safety Officer at the TUC where he has worked for over 11 years. His work involves research and policy development on a wide range of issues involving health and safety at work, including EC health and safety directives and their implementation in Britain. He is the author of the TUC guide *Tackling Stress at Work* and the co-author of the TUC's *No Excuses: Beat Bullying at Work.*

Brigid Proctor is now in the fourth stage of retirement. Previously a counsellor, trainer, supervisor and consultant, she is a Fellow and Accredited Supervisor of the British Association for Counselling. She was the co-founder of Cascade Training Associates and has published books, audiotapes and videotapes and Open Learning Materials on counselling and supervision.

Steve Rains is a Fellow of the Institute of Personnel and Development and a representative on the southern region of the Confederation of British Industry Regional Council. In 1992 he was appointed Employee Relations Manager for the south central division of the Royal Mail, based in Reading. In 1996 he was appointed Director of Communications for the division and became Director of Personnel in 1997.

Michael Scott works as an independent cognitive-behavioural counsellor in Liverpool. He is the author of *A Cognitive-Behavioural Approach to Clients' Problems* (Routledge, 1989), *Counselling for Posttraumatic Stress Disorder* (with Stradling: Sage, 1992, 2000), *Developing Cognitive-Behavioural Counselling* (with Stradling and Dryden: Sage, 1995), and *Brief Group Counselling* (with Stradling: Wiley, 1998).

Steven Stradling is a reader in Behavioural Aspects of Transport at the Transport Research Institute, Napier University, Edinburgh, where he undertakes research in traffic and transport psychology. He was previously a Senior Lecturer in the Department of Psychology at the University of Manchester University. As well as collaborating with Mike Scott he has also authored *Dealing with Stress* (with Thompson and Murphy: Macmillan, 1994) and *Meeting the Stress Challenge* (with Thompson, Murphy and O'Neill: Russell House, 1996, 1998).

Noreen Tehrani is an organisational counselling psychologist who is an authority on dealing with bullying in the workplace, having designed and introduced organisational approaches to assessing and responding to bullying at work. Her research and publications are in bullying, trauma and counselling.

Vivienne Walker is Head of Human Resources at the South and East Belfast HSS Trust. She has a strong interest in diversity in the workplace and has developed a number of initiatives to improve organisation approaches. Her most recent work on dignity in the workplace includes the development of a harassment support service that supports staff on a number of fronts including bullying. She has just developed a new service for staff in the South and East Belfast HSS Trust who are experiencing domestic abuse.

Foreword

Bullying at work has been emerging as a major problem for some time. It has significant adverse effects on individuals and on business success. We all, as members of society, have a duty to tackle the problem.

This book makes a valuable contribution to the task we all face by pulling together a number of different strands of thought and action.

Noreen Tehrani has gathered a wide range of contributors, respected in their particular fields, to provide insights and practical advice for everyone involved. There is information here for the victims of bullying and their colleagues at work, for the professionals involved and for the 'movers' and 'shakers' of our society who must ensure we all have the tools to help eradicate the problem.

Lord Monkswell
Chair of 'Campaign against Bullying at Work'

Preface

Over the past decade bullying in the workplace has been recognised as a problem facing many employees. Despite the high incidence of bullying in the workplace there is a conspicuous absence of published material on why these behaviours occur, how their occurrence can be reduced and what can be done to help the victims.

There is agreement between academics and practitioners that:

- organisational cultures can promote or undermine the dignity of the individual workers and influence their behaviour with respect to bullying;
- levels of bullying in the workplace are frequently underestimated;
- bullying causes significant health problems for employees in many organisations;
- even the threat of being bullied can be the cause of stress and ill health;
- building a culture of respect requires organisations to establish a range of integrated policies, structures and interventions;
- there is a need for greater understanding of the issues by a range of stakeholders including organisational management, unions, human resources, lawyers, general practitioners, occupational health psychologists and counsellors.

This book directly addresses these issues and provides a comprehensive review of the area together with a guide to taking effective actions. The book is divided into four inter-related parts.

Part 1 sets the scene by focusing on individual employees, the origins of bullying and the way that bullying is expressed in the workplace.

In Chapter 1 Helge Hoel and Cary Cooper describe workplace bullying and show how bullying affects individuals, groups, organisations and society. It is suggested that the causes of bullying are diverse and therefore it is unlikely that a single model or approach could adequately address the complex nature of the problem. In Chapter 2 Neil Crawford examines organisational responses to bullying. Drawing on personal experiences working in organisations and with individuals, he presents bullying as a symptom of organisational dysfunction. Chapter 3 by Michael Scott and Stephen Stradling presents their view that the effects of bullying on the individual are similar to those used in the diagnosis of post-traumatic stress, and that bullying is as distressing to the victim as being exposed to a physical attack or injury. In Chapter 4 Noreen Tehrani illustrates the effects on four people who had been exposed to bullying. Using debriefing and narrative techniques the employees were helped to describe in minute detail what happened during the bullying episodes, and the benefit of using debriefing and narrative therapy approaches in treating victims of bullying is highlighted in the process.

Part 2 looks at the difficulties that face organisations when trying to deal with bullying in the workplace. In Chapter 5 Claire Lawrence examines bullying in the context of social interaction. She looks at the extent to which the theories of social learning and modelling can be used as a framework for understanding the ways in which the individual can develop the use of bullying strategies as a way of obtaining rewards. In Chapter 6 Diane Beale looks at bullying from an organisational point of view, describing the range of terms and behaviours that have come to be considered as work-related bullying. Following a brief outline of a health and safety approach to tackling bullying, the chapter describes ways that the problem might be monitored effectively.

Part 3 looks at the legal requirements for organisations to take a proactive approach to reducing the likelihood of bullying occurring at work. In Chapter 7 Patricia Leighton describes how the law has come to recognise and to respond to bullying in the workplace. The chapter begins with an overview of the legal issues and describes how the tendency to focus on the identification and naming of specific forms of bullying has led to a failure to acknowledge and explore the nature of the underlying interpersonal behaviours found in bullying. The chapter presents challenging and refreshing views on the law.

Part 4 explores how organisations can proactively create an environment in which the organisation and the individual employees can show respect to one another.

In Chapter 8 Vivienne Walker looks at how an organisation's culture, standards and values impact on an employee's perception of what is acceptable behaviour. In her work in a health trust, Vivienne discusses how her organisation introduced a dignity at work policy that created a framework for addressing bullying and harassment issues. In Chapter 9 Noreen Tehrani looks at building a culture of respect through the use of a total quality approach. The advantages of using total quality as a way of changing organisational culture is built upon the well-tried and tested processes and systems that have been used with success within many organisations. It recognises that there is a need to include all the stakeholders in the process and that this universal participation increases the likelihood of success. In Chapter 10 Steve Rains provides a very practical account of how the Royal Mail introduced a peer 'listener' scheme. The issues involved in developing peer support programmes are discussed and the benefits of the scheme described.

In Chapter 11 Brigid Proctor and Noreen Tehrani look at the need to provide practical and emotional support for people in conflict. The need to protect supporters and counsellors from the impact of this work is described including the difficulty of providing support without getting trapped into the 'drama' of the bullying triangle that is often played out in the counselling room. In Chapter 12 Chris Manning provides a helpful chapter describing the issues that face a general practitioner dealing with patients affected by workplace bullying and suggesting how the victim of bullying can prepare for their first GP consultation.

The need to avoid over-medicalisation of bullying is emphasised as a way of ensuring that the symptoms and signs are recognised without disempowering the individual in the process. In Chapter 13 Tom Mellish illustrates through case studies how an organisational climate can be created in which employees are treated with dignity and respect through co-operation between the unions, workforce and management.

Finally in Chapter 14 Noreen Tehrani looks towards a future in which organisations are actively involved in creating working environments where bullying is no longer an acceptable way of behaving and in which employees and organisations treat one another with dignity and respect.

Acknowledgements

Categories of bullying behaviour (reproduced on p. 80 as Table 6.1) is from L. Quine's 'Workplace bullying in NHS community trust: staff questionnaire survey', *British Medical Journal*, 318, 7178, 1999, 228–32. It is reprinted by kind permission of BMJ Publishing Group.

The drama triangle (reproduced on p. 175 as Figure 11.1) was developed by Stephen B. Karpman, MD, and first presented in his article, 'Fairy tales and script drama analysis', published in the *Transactional Analysis Bulletin*, 7, 26, 1968, 39–43. It is reprinted here with the permission of the author and the International Transactional Analysis Association.

PART ONE

Origins and problems

CHAPTER ONE

Origins of bullying

Theoretical frameworks for explaining workplace bullying

HELGE HOEL AND CARY L. COOPER

Introduction

A growing body of evidence suggests that the problem of workplace bullying affects a substantial number of British employees either directly as targets of bullying behaviour or, indirectly, as bystanders or witnesses of such activities (Hoel and Cooper, 2000b). In order to be able successfully to prevent or at least reduce the prevalence of the problem, we need to have a clear view of the factors that cause it. Therefore, in this chapter we intend to trace the causes of bullying.

Much of the early UK literature on the problem of workplace bullying located the source of bullying in the personality of those involved, perpetrators as well as targets (Crawford, 1992; Field, 1996; Randall, 1997). By contrast, Scandinavian and German researchers primarily attributed the phenomenon of bullying to the work environment in which it occurs (Leymann, 1996; Vartia, 1996) or to an interplay or dynamic interaction of factors of a personal as well as a situational nature (Zapf et al., 1996; Einarsen, 1999). Recently attempts have also been made to bring society at large and the factors of an economic, social and political nature into the debate (Lee, 1998; Sheehan, 1999).

To reveal the origin of the bullying phenomenon in its many facets, we will explore some of the different theoretical approaches and perspectives that, so far, have been advanced. On the face of it, these approaches may appear to be competing and mutually exclusive. However, if taken as theoretical approaches or perspectives aimed at different levels of analysis (Einarsen, 1996; Hoel et al., 1999), these approaches may in fact complement each other, each serving a separate function depending upon which part of the problem we are exploring. We will, therefore, search for sources of bullying by means of an examination of the problem at the individual level, the dyadic interaction level, the group level, the organisational level and the societal level.

In order to make sense of the investigation consideration must also be given to issues related to definition and problem demarcation. With respect to demarcation, where we decide to draw the limits of the problem would impinge upon the breadth and focus of the subsequent investigation.

Defining workplace bullying

For the purpose of this chapter we will adopt a widely used definition by Einarsen et al. (1994b: 20):

> Bullying emerges when one or several individuals persistently over a period of time perceive themselves to be on the receiving end of negative actions from one or several persons, in a situation where the target of bullying has difficulty in defending him or herself against these actions.

This definition contains a number of features in need of further investigation.

Negative acts

We can divide the negative behaviour most frequently identified with bullying into the following categories: personal derogation (humiliation and personal criticism), work-related harassment (withholding of information and having one's responsibilities removed), social exclusion, violent threats and intimidation and work overload (Einarsen and Raknes, 1997; Hoel and Cooper, 2000b).

Persistent

Bullying will normally refer to behaviour that is repeated and persistent. Therefore, while it can be unpleasant to be the target of someone's occasional aggressive behaviour, such behaviour would in most cases be excluded from the definition. The exception here would be those cases where intimidating behaviour of a severe nature (e.g. physical violence or threat of physical violence) leaves the target in a permanent state of fear.

Long term

In some cases workplace bullying may be resolved in its early phases by means of organisational intervention or by initiatives from those involved or other concerned parties. In other incidents, one of the parties involved may decide to leave. However, the long-term nature of the phenomenon is one of the most salient features of the problem. In a recent study of 5,300 British employees, two out of three people currently bullied reported that the process had lasted for more than one year (Hoel and Cooper, 2000b). The prolonged nature of the exposure is particularly necessary in understanding the severe effects bullying may have on targets as well as the likelihood of finding a solution to the problem (Hoel and Cooper, 2000a).

Imbalance of power

When an equal balance of power exists between two individuals in a conflict situation we would not refer to the situation as bullying. However, it is important to note that power may be formal, drawn from one's hierarchical position within the organisation, or informal, referring to sources of power such as personal contacts, organisational standing and experience. Knowledge of an opponent's vulnerabilities may be a further source of informal power often exploited in cases of bullying.

Intent

Though central to many definitions of school bullying (e.g. Besag, 1989), intent is left out of the above definition. The role of intent, however, has given rise to considerable discussion and controversy within the bullying field (e.g. Keashly, 1998). There is disagreement as to the usefulness of including intent among defining characteristics, for example, whether bullying behaviour is necessarily used to harm a target or, at times, may be considered instrumental behaviour, where harm can be considered a non-intended side-effect of the behaviour. Owing to the difficulty of establishing intent in the context of litigation, it has been excluded as a defining feature in cases of other types of harassment that may be considered similar to bullying, e.g. sexual and racial harassment (Hoel, Rayner and Cooper, 1999). As will be evident from the subsequent exploration of the phenomenon, this does not imply that the role of intent is unimportant in understanding bullying and the victimisation process. In fact, according to Einarsen (1999) the attributions made by targets may have as serious an effect on the victim as the negative behaviour itself. (For a more comprehensive discussion, see Hoel, Rayner and Cooper, 1999.)

Subjective perception

Most researchers today would agree that what matters in cases of bullying is the subjective perceptions of targets, or the meaning they attach to their experience. In other words, to understand the phenomenon of bullying we need to explore those factors that may influence how the negative behaviour or bullying acts are perceived (Einarsen, 1999).

Problem demarcation

In media accounts of workplace bullying in Britain, bullying is frequently identified with negative behaviour *per se*. However, in line with interpretations shared by many researchers and practitioners in the field (e.g. Einarsen, 1999), we consider bullying to be a dynamic phenomenon, where the presence of negative behaviour is seen as a necessary part of the problem but not sufficient to explain the phenomenon in its entirety. Thus, the phenomenon of bullying is seen to be unequivocally tied in with the conflict escalation process in which the response of the targets is seen as integral with process development and outcomes.

At the individual level

In trying to explain workplace bullying by reference to the make-up and personality of those involved, simple popular beliefs often overlap and blend in with other more reflective and theoretically based views. This is not least the case where bullying is explained by reference to pathological and deviant personality traits identified with those of a psychopath (Field, 1996). The British media picture of the adult bully is also one that may fit well with our understanding of psychopathic behaviour.

The interest in the role of personality traits in explaining workplace bullying has largely been informed by research into school bullying. Arising from this research, bullies are recognised as being identified with high levels of aggressiveness. In contrast victims appear to be anxious and to have low self-esteem (Olweus, 1997). These findings have predominantly been explained with reference to early childhood experiences. According to Randall (1997) both aggressive as well as submissive personalities may be seen to result from poor or unsuccessful parenting. As far as the aggressors or bullies are concerned, they appear to come from home environments where inconsistent patterns of discipline, authoritarianism or even rejection on the part of parents are seen as normal. Not only are such parenting styles seen to stimulate aggression, their presence appears to reduce the opportunity for the development of social skills. For victims of bullying, parental rejection may lead to timid, submissive and over-protected children. Being over-protected by parent-figures in childhood is also considered to undermine the development of social skills, a characteristic which victims appear to share with bullies. On a similar note, a feature identified with many bullies and victims is their apparent lack of necessary problem-solving skills (Perry et al., 1998).

In order to explain the behaviour of bullies and victims alike, the 'social learning theory' of aggression (Bandura, 1973) is frequently applied (Randall, 1997). According to social learning theory, bullying can be seen as a variant of repeated aggression, and be understood as a learned set of behaviour, primarily stimulated by external sources or modelling (Bandura, 1977). By successfully applying aggression (in other words, getting what you want), such positive reinforcement makes the behaviour more likely to be repeated in the future. The intent behind the aggression or bullying should also be seen as being connected with the expectation of particular outcomes. Thus, according to Randall (1997), expectation of victim distress is part and parcel of the bullying acts. It follows that bullies will pick their victims with great care in order to ensure a successful outcome and positive reinforcement of their behaviour. Therefore, victims who are easily brought to submission appear to make perfect targets in the eyes of the bully (Randall, 1997).

According to the social learning theory of aggression, personality traits such as aggression and submissiveness can, when established, be considered to be relatively stable over time. This gives rise to the idea of a 'cycle of violence' (Tattum and Tattum, 1996; Randall, 1997), suggesting that bullying behaviour, established and perfected in childhood and adolescence, continues to express and manifest itself in a variety of situations throughout life (Randall, 1997). In support of such a 'cycle of violence', results of some longitudinal studies of childhood aggression are often quoted (e.g. Eron et al., 1987).

However, with reference to the likely negative consequence of being 'labelled' at an early stage, Hoel et al. (1999) warn against reading too much into these findings. The fact that many people, who have a past as either bully or target of bullying at school, apparently succeed both professionally and personally later in life suggests that matters are more complicated. Thus recent developments in aggression theory tend to emphasise the role of situational factors in evoking aggressive behaviour (Geen, 1990; Neumann and Baron, 1998).

The evidence of the presence of particular personality traits among perpetrators and targets of workplace bullying is at present sparse. Owing to the difficulty in researching the bullying personality, the evidence that has been produced is predominantly linked to targets. For example low self-esteem and higher than normal anxiety levels are among personality traits frequently identified with targets (Einarsen et al., 1994). In a recent study, individuals who were introverted, conscientious, neurotic and submissive were found to be more likely to be targeted for bullying and subsequently victimised (Seigne et al., 1999).

To state that anxiety, neurotic behaviour and low self-esteem are the predominant personality characteristics of many victims is uncontroversial. What is causing debate, however, is whether, and to what extent, these personality characteristics actually should be considered causes of bullying or whether they are the end-product or the result of the bullying process. With reference to the neurotic and often obsessive behaviour of many victims of bullying, Leymann (1996) argues that their behaviour needs to be understood as a normal response to an abnormal situation. Similarly, when one sees one's own world collapsing as a result of exposure to bullying, it would not be surprising if the experience affected the way an individual sees the world and their ability to influence and control events (Hoel and Cooper, 2000a). However, we would tend to agree with Einarsen et al. (1994b) who state that bullying is 'neither the product of chance nor of destiny. Instead, bullying should be understood primarily as an interplay between people, where neither situational nor personal factors entirely suffice to explain why bullying develops' (translated by Hoel, 1997). However, before we consider the more dynamic aspects of the bullying process as understood in connection with dyadic interaction, we need to examine some other factors which tend to play a role in cases of bullying and operate on the level of the perceptions of individuals.

The role of attribution processes in the bullying process is one such factor that has been emphasised by a number of researchers (Niedl, 1996; Hoel et al., 1999). According to attribution theory (Kelly, 1972), individuals tend to project positive experiences towards themselves, while behaviour or acts that may be considered negative are projected on to others. It is therefore not surprising that targets of bullying tend to attribute blame to external sources or to the work environment, as well as to opponents, as opposed to blaming themselves. Similarly, people tend to explain their own behaviour by pointing to the environment and the circumstances in which the behaviour occurs, while other people's behaviour is explained as a product of their individual personality. This phenomenon is often referred to as the 'fundamental attribution failure' (Jones and Davies, 1965). From studies of targets of bullying, it has become evident that as the conflict progresses, targets will increasingly attribute blame to their opponent while simultaneously attributing less responsibility for events to themselves (Kile, 1990). In order to establish a picture of events with any degree of certainty, such attribution processes must be considered during any investigation (Hoel et al., 1999).

The role of individual perception is also emphasised in the approach put forward by Liefooghe and Olafsson (1999). They argue that, as bullying is a 'new', or newly acknowledged social problem, the perceptions of it are likely to vary between individuals. In their view, faced with negative behaviour, people draw on what they refer to as 'social representations' in order to make sense of the behaviour to which they are being exposed. Social representations are interpretative frameworks that may account for an event or a phenomenon. When a particular event or behaviour is considered, a number of aspects concerning the event or the phenomenon, for example explanatory models, likely effects, etc. immediately spring to mind. As a newly 'discovered' phenomenon bullying may correspond to a number of social representations and the way individuals may feel about being exposed to bullying may vary according to the social representations they invoke. Furthermore, the activation of these social representations is an active process where the individual may consider a number of social representations to account for an event. As the outcome of these considerations or individual negotiations will decide how we will feel and subsequently act, it is important to study these processes in more detail. Moreover, to the extent that perceptions may be shared amongst groups, identification of such shared perceptions may become a powerful tool in addressing the problem locally.

At the dyadic level

So far we have sought the origin of bullying in the personalities of those involved and in the cognitive processes that may inform their behaviour. We will now move on to identify possible sources of bullying in the interaction between perpetrator and recipient or, more correctly, between the two main parties involved in the unfolding process.

In most cases of bullying the target is not entirely a passive recipient of negative acts and behaviours. Each negative act delivered is likely to produce a response on the part of the recipient, which again is likely to impinge upon the further response of the aggressor. It is this particular pattern of action and reaction as it unfolds in cases of bullying which is at the centre of a transactional perspective or theory of bullying (Einarsen, 1999). These dynamics will be influenced by the personalities of those involved. In the same way as low self-esteem may produce a particular response, unrealistically high self-esteem is also likely to affect the process (Matthiesen et al., 1999). The ability to see situations from the perspective of others may play a part here. Thus, depending on one's response to a negative act, one may succeed in neutralising or deflecting the situation or the response may cause the conflict to escalate. It is important to emphasise that responses, which with hindsight may be considered to have contributed to escalating the process, may have appeared rational and sensible at the time of response, even from a conflict-resolution perspective.

In a case study of bullying, exploring the process from the perception of the victim, Matthiesen et al. (1999) throw some light on the complexity of the interaction process. They demonstrate how a particular course of action on the part of the target turned out to have very different consequences than was first intended by that target. In other words, acts and behaviours that were intended to reduce and even solve the conflict as far as the target was concerned turned out to influence the preconditions of the negative behaviour, thereby further escalating the process. For example, in one of the cases explored, the interviewee experienced her work colleagues turning against her as a result of her seeming ease of access to professional and legal expertise when challenging the conduct of her employer.

In a recent article Einarsen (1999), referring to the work of Felson and Tedeschi (1993), introduces a distinction between two forms of bullying: predatory and dispute-related bullying. Predatory bullying refers to those situations where an individual is targeted for negative behaviour or abuse by accident or by chance. In other words, no previous provocative behaviour on the part of the recipient can account for the behaviour the target experiences. This is the type of bullying encountered in many of the high-profile cases recently brought to court in Britain. According to Einarsen (1999), predatory bullying is often identified with a demonstration or public manifestation of power, and is frequently linked to the culture of the organisation. The fact that it is purely fortuitous whether one may be targeted or not does not necessarily imply that everybody has an equal chance of becoming a target of bullying. In such circumstances it may be sufficient to be defined as belonging to a group of outsiders or to manifest certain characteristics which may draw attention to oneself (Einarsen, 1999).

By contrast, dispute-related bullying refers to those incidents of bullying which are connected with disagreement over or violation of social norms. If such conflicts are not resolved they may gradually escalate, turning increasingly personal. At this stage of conflict escalation the main aim of the players appears to be to grind the opponent down. It is worth noting that, at the early stages of conflict escalation, it may even at times be difficult to tell who is going to gain the upper hand, especially when the conflict is between peers or colleagues. This is particularly so since either of the opponents may adopt the stance of victim. In such a situation, claiming victim status and accusing one's opponent of bullying may be powerful tactics in such a power struggle (Einarsen, 2000). However, when the power struggle has reached its peak, victimisation of the weaker of the opponents is a likely outcome. For the observer as well as for the parties involved, the origin of the conflict may at this stage be unclear as the reason for the dispute may change during the course of events.

At the group level

Bullying can also be explained with reference to'scapegoating' processes within workgroups and organisations (Thylefors, 1987; Crawford, 1997). The origins of the scapegoat metaphor are in the Old Testament, Leviticus 16, where a goat was sent out into the wilderness after the Jewish chief priest had symbolically laid upon it the sins of the people. Historically witch-hunts may be considered to be examples of such collective scapegoating processes. This approach is primarily concerned with psychodynamic factors or characteristics of inherent human interactions within workgroups and organisations.

In the psychodynamic approach expression and the disposal of aggression (e.g. interpersonal conflict) are deemed to be normal features of everyday life. Thylefors (1987) argues that bullying is largely to be understood as a process whereby the identification of scapegoats fulfils certain personal and organisational needs. Our personality and personal characteristics are seen predominantly as products of childhood experiences in which the individual, faced with problems and frustrations, has developed necessary coping strategies. Our ability to co-operate and to relate to others seems to reflect the extent to which these coping strategies have been adjusted or compensated for through later experiences and influences. However, under stress and in particular situations that resemble experiences in childhood, old strategies and behavioural patterns might be rediscovered and applied to the present situation (Hoel, 1997). Ambiguous situations, or situations where the real source of frustration is unclear, also represent fertile ground for scapegoating processes. Furthermore, when the real source of conflict appears to be hidden from us, or considered to be out of reach or impossible to influence, scapegoating may be a likely outcome (Einarsen, 2000).

According to Thylefors (1987), typical behaviour that may contribute to attaining the status of scapegoat includes being too honest, showing lack of willingness to compromise and 'anachronistic' behaviour. This last type refers to behaviour that does not keep pace with development within the group and organisation. Prejudice is another factor that may influence the choice of target. Being part of a minority or an outsider group may also affect whether aggressive acts are performed or initiated (Einarsen, 2000). By projecting our frustration on to people who are considered weak or unlikely to retaliate we are also more likely to get away with our behaviour (Einarsen, 2000).

Schuster (1996) suggests that a clear distinction needs to be made between dyadic bullying, on the one hand, and those incidents of workplace bullying which involve several perpetrators or even an entire workgroup, on the other. In support of her view she refers to the work of Coie and Dodge (1988) on 'peer-rejection' in a school context. According to their research, socially-rejected children have a tendency to attribute hostile intentions to the behaviour of others and to lack social competence. According to Schuster (1996), social incompetence may often reveal itself at the stage of group entry where failure to read informal group rules may translate itself into rapid rejection. Rule violation may also evoke stricter rule compliance by the rest of the group. Rejected children were also identified with less pro-social behaviour and were more likely to act aggressively than other children (Coie and Dodge, 1988). However, as far as aggressiveness is concerned, it is not aggressiveness *per se* which appears to evoke rejection, but what may be referred to as 'unprovoked aggression' or where there seems to be no apparent reason for aggressive behaviour (Coie et al., 1990). In the literature on school bullying such behaviour is often identified with that of 'provocative victims'. While we should be careful about transferring knowledge from a school setting to the world of work, by studying the processes of rejection and isolation as they may apply to a work setting, we may be able to extend our knowledge of bullying further. However, there is a thin line between exploring the problem, on the one hand, and blaming the victim, on the other, and we need to ensure that we stay on the right side of this line.

At the organisational level

In order to explain and identify the origins of bullying within the context of the organisation, various models (which often relate to the concept of stress), have been put forward (Zapf et al., 1996). In this respect German and Austrian researchers appear to go furthest, referring to bullying as a severe variant or subset of social stress and an indicator of workplace conditions (Niedl, 1995). The focus of these approaches has been on the likely antecedents of bullying within the context of the organisation. As antecedents of a personal nature have already been considered at length, we will focus on the situational and cultural antecedents of bullying.

Research on the social antecedents of bullying has mainly focused on the role of the organisational norm and the relationship between the bullying and socialisation processes at work. In his study of bullying in the Fire Service, Archer (1999) explores how bullying may become institutionalised and passed on as tradition. Archer identifies the training process as a powerful source of institutionalisation of behaviour, in particular where every uniformed member of the organisation shares the same experience. With regard to behaviour that may be construed as bullying, it was found that victims and onlookers alike reported that the perpetrators had been exposed to a similar experience sometime in their own career. Furthermore, organisational members, including victims, interpreted the negative treatment experienced by targets of bullying, as customary behaviour and not as vindictive behaviour aimed at harming the individual as such. Situational factors, such as the 'watch' culture, where the individual is allocated to the same tightly knit work-team, possibly for years at a time, also suggest little room for diversity. Moreover, in an autocratic leadership culture where one's superior has been brought up within the same tradition, it is difficult to break out of the cycle and embark upon cultural change. The fact that many victims considered complaining about bullying to be an act of disloyalty further emphasises the potential strength and impact of the socialisation processes at work (Archer, 1999).

Several Scandinavian studies of workplace bullying have explored situational antecedents of bullying (Einarsen et al., 1994a; Vartia, 1996). According to Einarsen et al. (1994a), workplaces that report a high number of incidences of bullying appear to be those with few challenges, less variety, and less interesting work. They also tend to present few possibilities for personal development, a less satisfactory social climate, less influence on one's own work situation and more dissatisfaction with superiors, as well as the prevalence of time-pressure. However, no organisational or industry-wide differences have so far been identified. More specifically their results suggest that among characteristics of the work-environment, 'leadership', 'role-conflict' and 'work-control' appear to have a strong bearing on bullying. As far as 'leadership' is concerned, it is lack of leadership or abuse of power that can potentially bring about bullying. According to Zapf et al. (1996) restricted control over time frequently undermines the opportunity for conflict resolution, which may leave potentially damaging conflicts unresolved and thus open for future escalation. In a similar way Vartia (1996) concludes that work autonomy, poor communication and lack of involvement are typical features of the work environment that promotes bullying (Hoel, 1997).

While some situational antecedents of bullying may be similar across occupations and jobs, others may be specific to the context in which they arise. A work environment that has frequently been identified with a high prevalence of bullying, often of a violent nature, is the restaurant kitchen. By studying workplace bullying within the kitchen context we may be able to expose a number of situational and social factors, which alone and in combination may interact with characteristics of the individuals involved to cause bullying behaviour and victimisation. Kitchen work is known to be hot and noisy, and at times very stressful. This is particularly true of the luxurious restaurants that appeal to the customer who is willing to spend a considerable sum of money on a meal on the assumption that one is being guaranteed a top-quality product. Over a frenetic few hours when the restaurant is filled to the last chair, customer satisfaction must be guaranteed, otherwise the chef's reputation will be at stake. The fact that many of today's chefs have a personal financial interest in the business is likely to exacerbate the pressure under which they operate (Johns and Mentzel, 1999).

The growing interest in preparing food and cooking in the media has not only provided considerable financial benefit for some chefs but also meant that a number of them have achieved personal stardom. According to Johns and Mentzel (1999) this has also given rise to the concept of the chef as an artist, whose bullying and abusive behaviour must be understood as idiosyncratic behaviour born out of their artistry and creativity. The media has also played an active part in creating this myth.

A typical example of such image-making is a newspaper article (*Guardian*, 19 August 1996) featuring a named top chef (introduced as the bad boy of London's restaurant world) and his culinary and economic success under the heading: 'The rise and rise of an abrasive chef, from *enfant crédible* to *enfant incrédible*'. For the young apprentices, who not only model themselves and their future on the creative and gastronomic qualities of their role-model, such behaviour may be considered normal and part and parcel of work as a chef. As was the case in the previous example of bullying in the Fire Service, a different cycle of abuse and bullying is emerging, where bullying is being institutionalised and handed down from one generation of chefs to the next. Moreover, in the case of chefs, the effect of their behaviour is not limited to the apprentice group, but would affect all kitchen staff in general.

At the societal level

It is impossible to give a comprehensive picture of the various causes of workplace bullying without acknowledging the impact of societal factors. While the particular antecedents of bullying will vary between organisations, most organisations at the beginning of the new millennium are still in the midst of struggling with the effects of significant change processes. In order to sustain competitiveness in a global marketplace, organisations are forced to embark upon significant organisational and technological change (Cooper, 1999). Shrinking profit margins and volatile markets give impetus to restructuring and downsizing, processes often undertaken in an aggressive manner (McCarthy et al., 1995). As a result, employees at all levels of the organisation find themselves facing increasing workloads, often in a climate of uncertainty with regard to their future employment situation (Stewart and Swaffield, 1997). The problem is exacerbated by the increase in working hours, which as far as Britain is concerned, are currently the longest in Europe (Worrall and Cooper, 1999), and a growing number of people holding temporary contracts and part-time jobs. With greater job insecurity, employees become less resistant to pressure and more unlikely to challenge unfair and aggressive treatment on the part of managers. Managers themselves are often required to work excessively long hours, with increased responsibility and accountability (Lewis, 1999), as the human resource function is now often devolved to the level of the line-manager (James, 1992). With managers being given greater opportunity to wield power in a situation where employees may be less likely to challenge any abuse, bullying may thrive.

Faced with this situation, and bearing in mind the compression of career structures resulting from delayering processes, itself an outcome of the drive for leaner organisations, it is not surprising to find that managers frequently make use of authoritarian and even abusive behaviour in order to carry out their work (Sheehan, 1999). Starved of time and resources, managers may similarly employ bullying tactics in order to 'get the job done'. As such, bullying may increasingly take on an instrumental character, where the bullying may be considered as a means to an end. As discussed in the previous section, external pressures are also likely to evoke aggressive reactions and responses from subordinates, increasing the likelihood of bullying scenarios developing. Moreover, the fact that high levels of bullying are being reported from top to bottom of the organisational hierarchy (Hoel and Cooper, 2000b), suggests that aggressive behaviour and bullying tactics are being cascaded downwards within the organisation. Thus, individuals at each level are increasingly likely to employ bullying tactics, possibly as a matter of survival in situations that, to the individual, may appear unsolvable by other means.

Another factor to consider is the changing nature of work in Britain in the 1990s and its effect on the employment relationship. By introducing market-place philosophies into areas previously not affected by such pressures, e.g. the National Health Service (NHS) and the education sector, the relationship between managers and employees has been changed (Lee, 1998). As a result management pressure on employees has increased significantly (NASUWT, 1996). According to Lee (1998) in this situation unfair treatment and harassment on the part of management are less likely to be challenged by employees.

The change in the employment relationship needs to be viewed in connection with other factors such as the weakening of the trade union movement and the growing individualisation of British society in general. As far as the trade unions are concerned, these changes are also reflected in trade union practices, where collective actions have frequently been replaced by individual grievances (Bacon and Storey, 1996). It follows that collective responses to abusive management behaviour may become less likely. As a result, employees may come to interpret behaviour in an individualised manner. Thus management behaviour, which previously may have been perceived within an 'us and them' framework, may now be interpreted as a personal attack and as such may take on connotations of shame and self-blame, making bullying and victimisation more likely outcomes.

Bullying may also be explained with reference to employee resistance to managerial control that could provoke retaliatory behaviour on the part of managers. Employee resistance can take many forms and may include behaviours and actions of a passive as well as an active nature, and be performed on an individual basis as well as in a more collective manner. US management literature has paid a lot of attention to such resistance under various labels such as 'organisational misbehaviour' (Vardi and Wiener, 1996) and 'Dysfunctional work behaviour' (Griffin et al., 1998). These labels refer to a wide set of behaviours from theft and sabotage, at one end of the spectrum, to withholding of effort and talking with co-workers instead of working, on the other. Interestingly enough, few references are made to abusive behaviour on the part of managers. Without questioning the background of any changes taking place and their likely impact, these approaches may function as justification for stronger managerial control, as well as deeming illegitimate any resistance to managerial control in general and to the restructuring processes in particular.

In a period of rapid social and economic change, government action and rhetoric may similarly contribute indirectly to workplace bullying. For example, with reference to the 'common good', government may argue that 'unions are unwilling to traverse the waters of change and present obstacles to policy shift' (Williams, 1994: 4). Such a description of the role of government may fit the situation in Britain well. In a very public attack the Labour government accused public-sector employees and trade unions of representing 'forces of conservatism' for their seeming unwillingness to embrace change, when opposing new government initiatives linked to attitudes to inspection systems and performance-related pay (*Guardian*, 22 October 1999). Such public attacks could be interpreted as indirect support for local management pressure to force through government policy in a manner that could be construed as bullying by those exposed to the pressures. This could be the case particularly when the resistance on the part of employees, e.g. teachers or health carers, is professionally and ideologically motivated with its basis in their own personal views, judgement and experience.

Discussion

In our exploration of the origins of workplace bullying, we have attempted to identify how different causes of workplace bullying on their own, or in combination, may produce bullying behaviour and victimisation. The theoretical perspectives and frameworks studied should not be understood as mutually exclusive. On the contrary, there are strong overlaps between them, with knowledge of one perspective often seen as a precondition for understanding another. For example, knowledge of the impact of personality factors is seen as essential to make sense of bullying across the differing levels of analysis. The same argument may be applied to the role of attribution processes. While we cannot rule out the possible influence of factors rooted in genetics and inherent human nature, it should be apparent from the previous discussion that we consider the impact of nurture to be the more important explanatory factor when it comes to workplace bullying. Moreover, when considering intervention at an organisational level, initiatives focusing on personality and personality characteristics are seen as unlikely to succeed (Rayner, 1999; Hoel and Cooper, 2000a). This is the case whether the aim is to change the person or prevent certain persons from gaining promotion or from entering the organisation altogether. This should not be taken as an argument against organisational initiatives aimed at behavioural change, however. Disagreement may also exist as to what level of analysis certain factors and processes should be assigned.

The theoretical perspective/s one decides to adopt also depend upon one's intentions and will be influenced by factors such as our organisational role and position, as well as which part of the bullying process and with what level of analysis one is concerned. It follows that a human resource practitioner may apply a different perspective from that of a counsellor or an occupational health practitioner, in the same way that the needs of a national officer of a trade union may differ from those of a local shop steward.

In this chapter we have endeavoured to highlight the complexity of the problem of workplace bullying. Simple explanations cannot do justice to this issue and are likely to obscure rather than facilitate understanding. To construct bullying as a problem identical with autocratic and abusive management styles, on the one hand, or the product of the behaviour of psychopaths, on the other, should therefore be avoided. Indeed, some authors, notably Einarsen (1999) have questioned altogether the notion of workplace bullying as a single unified problem, suggesting instead that it is better to explain bullying as a host of problems, where 'bullying' can be understood as an umbrella term. Following this view, Rayner (1999) suggests that we may need to extend our vocabulary to account for the different aspects and manifestations of bullying. Moreover, we have emphasised that bullying is a dynamic interactive process with a multiplicity of causes, where possible process outcomes may serve as antecedents for further conflict escalation (Hoel et al., 1999). This has important implications for both researchers and practitioners. Before deciding which theoretical explanation or framework to apply, it is necessary to establish what part of the problem-conglomerate one intends to investigate.

References

Archer, D. (1999) 'Exploring "bullying" culture in the para-military organisation', *International Journal of Manpower*, 20, 94–105.

Bacon, N. and Storey, J. (1996) 'Individualism and collectivism and the changing role of trade unions' in P. Ackers, C. Smith and P. Smith (eds) *The New Workplace and Trade Unionism*, London: Routledge.

Bandura, A. (1973) *Aggression: A Social Learning Analysis*, Englewoood Cliffs, NJ: Prentice-Hall.

Bandura, A. (1977) *Social Learning Theory*, Englewood Cliffs, NJ: Prentice-Hall.

Besag, V. (1989) *Bullies and Victims in Schools*, Milton Keynes: Open University Press.

Coie, J. and Dodge, K. A. (1988) 'Multiple sources of data on social behaviour and social status in the school: a cross-age comparison', *Child Development*, 59, 815–29.

Coie, J., Dodge, K. A. and Kupersmidt, J. B. (1990) 'Peer group behaviour and social status' in S. R. Asher and J. D. Coie (eds) *Peer Rejection in Childhood*, New York: Cambridge University Press, 17–59.

Cooper, C. L. (1999) 'The changing psychological contract at work', *European Business Journal*, 11, 115–18.

Crawford, N. (1992) 'The psychology of the bully' in A. Adams (ed.) *Bullying at Work: How to Confront and Overcome It*, London: Virago.

Crawford, N. (1997) 'Bullying at work: a psychoanalytical perspective', *Journal of Community and Applied Social Psychology*, 7, 219–25.
Einarsen, S. (1996) *Bullying and Harassment at Work: Epidemiological and Psychological Aspects*, PhD thesis, Department of Psychological Science, University of Bergen.
Einarsen, S. (1999) 'The nature and causes of bullying at work', *International Journal of Manpower*, 20, 1 and 2, 16–27.
Einarsen, S. (2000) 'Mobbing i arbeidslivet: hva, hvem, hvordan og hvorfor?' in S. Einarsen and A. Skogstad (eds) *Det Gode Arbeidsmiljø: Krav og Utfordringer*, Bergen: Fagbokforlaget, 167–82.
Einarsen, S. and Raknes, B. I. (1997) 'Harassment in the workplace and the victimization of men', *Violence and Victims*, 12, 247–63.
Einarsen, S., Raknes, B. I. and Matthiesen, S. B. (1994a) 'Bullying and harassment at work and their relationships to work environment quality: an exploratory study', *European Work and Organizational Psychologist*, 4, 4, 381–401.
Einarsen, S., Raknes, B. I., Matthiesen, S. B. and Hellesøy, O. H. (1994b) *Mobbing og Harde Personkonflikter: Helsefarlig Samspill på Arbeidsplassen*, Oslo: Sigma Forlag.
Eron, L. D., Huesmann, E., Romanoff, R. and Yarnel, P. W. (1987) 'Childhood aggression and its correlates over 22 years' in D. H. Gowell, I. M. Evans and C. R. O'Donnel (eds) *Childhood Aggression and Violence*, New York: Plenum.
Felson, R. B. and Tedeschi, J. T. (1993) *Aggression and Violence: Social Interactionist Perspectives*, Washington DC: American Psychological Association.
Field, T. (1996) *Bully in Sight: How to Predict, Resist, Challenge and Combat Workplace Bullying*, Wessex Press: Vantage, Oxfordshire.
Geen, R. G. (1990) *Human Aggression*, Buckingham: Open University Press.
Griffin, R. W., O'Leary-Kelly, A. and Collins, J. (1998) 'Dysfunctional work behaviours in organizations' in C. L. Cooper and D. M. Rousseau (eds) *Trends in Organizational Behaviour*, 5, 66–82.
Hoel, H. (1997) 'Bullying at work: a Scandinavian perspective', *Institution of Occupational Safety and Health Journal*, 1, 51–9.
Hoel, H. and Cooper, C. L. (2000a) 'Working with victims of workplace bullying' in H. Kemshall and J. Pritchard (eds) *Good Practice in Working with Victims of Violence*, London: Jessica Kingsley Publishers, 101–18.
Hoel, H. and Cooper, C. L. (2000b) 'Destructive conflict and bullying at work', unpublished report.
Hoel, H., Rayner, C. and Cooper, C. L. (1999) 'Workplace bullying' in C. L. Cooper and I. T. Robertson (eds) *International Review of Industrial and Organizational Psychology*, 14, 195–230.
James, P. (1992) 'Reforming British Health and Safety Law: a framework for discussion', *Industrial Law Journal*, 21, 83–105.
Johns, N. and Menzel, P. J. (1999) 'If you can't stand the heat!: kitchen violence and culinary art', *Hospitality Management*, 18, 99–109.
Jones, E. E. and Davis, K. E. (1965) 'From acts to dispositions: the attribution process in person perception' in I. L. Berkowitz (ed.) *Advances in Experimental Social Psychology*, 2, New York: Academic Press.
Keashly, L. (1998) 'Emotional abuse in the workplace: conceptual and empirical issues', *Journal of Emotional Abuse*, 1, 85–117.
Kelly, H. H. (1972) 'Attribution in social interaction' in E. E. Jones (ed.) *Attribution: Perceiving the Causes of Behaviour*, Morristown, NJ: General Learning Press.

Kile, S. M. (1990) *Helsefarleg Leiarskap: Ein Eksplorerande Studie*, rapport til Norges Almenvitenskapelige Forskningsråd, Bergen.

Lee, D. (1998) 'The social construction of workplace bullying', unpublished paper.

Lewis, D. (1999) 'Workplace bullying: interim findings of a study in further and higher education in Wales', *International Journal of Manpower*, 20, 1 and 2, 1, 106–18.

Leymann, H. (1996) 'The content and development of mobbing at work', *European Journal of Work and Organizational Psychology*, 5, 2, 165–84.

Liefooghe, A. P. D. and Olafsson, R. (1999) '"Scientists" and "amateurs": mapping the bullying domain', *International Journal of Manpower*, 20, 39–49.

Matthiesen, S. B., Aasen, A. and Olsnes, G. (1999) *Case Studie: Mobbing i Jobben Belyst Gjennom Bruk av en Konfliktmodell*, Institutt for samfunnspsykologi, University of Bergen: Bergen.

McCarthy, P., Sheehan, M. and Kearns, D. (1995) *Managerial Styles and their Effects on Employees' Health and Well-being in Organizations Undergoing Restructuring*, Brisbane: School of Organizational Behaviour and Human Resource Management, Griffith University.

NASUWT (1996) *No Place to Hide: Confronting Workplace Bullies*, Birmingham: NASUWT.

Niedl, K. (1995) *Mobbing/bullying am Arbeitsplatz*, Munich: Rainer Hampp Verlag.

Niedl, K. (1996) 'Mobbing and well-being: economic and personnel development implications', *European Journal of Work and Organizational Psychology*, 5, 203–14.

Neumann, J. H. and Baron, R. A. (1998) 'Workplace violence and workplace aggression: evidence concerning specific forms, potential causes, and preferred targets', *Journal of Management*, 24, 391–412.

Olweus, D. (1997) 'Bully/victim problems in school: knowledge base and an effective intervention program', *The Irish Journal of Psychology*, 18, 170–90.

Perry, D. G., Kusel, S. J. and Perry, L. C. (1988) 'Victims of peer aggression', *Developmental Psychology*, 24, 807–14.

Randall, P. (1997) *Adult Bullying: Perpetrators and Victims*, London: Routledge.

Rayner, C. (1999) 'Theoretical approaches to the study of bullying at work', *International Journal of Manpower*, 20, 10–15.

Schuster, B. (1996) 'Rejection, exclusion, and harassment at work and in schools', *European Psychologist*, 1, 293–317.

Seigne, E., Coyne, I. and Randall, P. (1999) 'Personality traits of the victims of workplace bullying: an Irish sample', Ninth European Congress of Work and Organizational Psychology, 12–15 May, Espoo, Finland.

Sheehan, M. (1999) 'Workplace bullying: responding with some emotional intelligence', *International Journal of Management*, 20, 57–69.

Stewart, M. B. and Swaffield, J. K. (1997) 'Constraints on the desired hours of work of British men', *Economic Journal*, 107, 520–35.

Tattum, D. and Tattum, E. (1996) 'Bullying: a whole school response' in P. McCarthy, M. Sheehan and W. Wilkie (eds) *Bullying: From Backyard to Boardroom*, Alexandria, Australia: Millennium Books, 13–23.

Thylefors, I. (1987) 'Syndbockar: om utstøtning och mobbning i arbeidslivet', *Natur och Kultur*, Stockholm.

Vardi, Y. and Wiener, Y. (1996) 'Misbehavior in organizations: a motivational framework', *Organizational Science*, 7, 151–65.

Vartia, M. (1996) 'The sources of bullying: psychological work environment and organisational climate', *European Journal of Work and Organizational Psychology*, 5, 203–14.

Williams, L. C. (1994) *Organizational Violence: Creating a Prescription for Change*, Westport, CT: Quorum Books.

Worrall, L. and Cooper, C. L. (1999) *The Quality of Working Life: 1999 Survey of Managers' Changing Experiences*, London: The Institute of Management.

Zapf, D., Knortz, C. and Kulla, M. (1996) 'On the relationship between mobbing factors, and job content, social work environment, and health outcomes', *European Journal of Work and Organizational Psychology*, 5, 215–37.

CHAPTER TWO

Organisational responses to workplace bullying

NEIL CRAWFORD

> At times I am not a reasonable man; you will forgive me for not going into the details here. Over the last 20 years, I have gone into organisations to try and understand with them how groups and individuals work together. The fallible, unreasonable part of myself has stood me in good stead and helped me to understand the irrational, the sadistic, the vengeful, and the aggression that can lie just below the surface of organisations festering one day, bursting out another.
>
> Neil Crawford (2000)

Introduction

In this chapter we will look at the organisational responses to bullying, for management must satisfy itself that reports of bullying are not symptoms of wider problems inside the organisation. Managers should view reports of bullying as opportunities for understanding, rather than as a drain on their time.

It may be difficult to take on board the fact that many people bring the legacy of their past to work and take it out on those they work with. By definition a personal problem becomes an organisational concern if it 'knocks-on' to others in the working group. It can be argued, of course, that it is not the aim of the organisation to understand itself. Every organisation has a task, and the people in it are primarily concerned with the fulfilment of that task. It does not exist to further the understanding of the workforce, except in relation to getting the job done. After all the livelihoods of all concerned depend on the company fulfilling its obligations.

This persuasive argument – that we concern ourselves with what we do not know we are – can seem convincing, except when the conflicts in human relations interfere with, or impede, the primary task. Policies then can only ever be a starting point with which to work with all these behaviours including workplace bullying.

However, it would be remiss if one didn't allude to childhood experiences that can determine adult responses. Adult patterns of behaviour are formed in childhood. Much of our behaviour at work springs from conflicts within ourselves which come from our early family life. Some situations encourage our maturity to flourish while others leave us floundering.

The motivation of those working in the field of bullying

One of the most important questions that a worker in the field of bullying needs to ask is why am I drawn to this? Interest in this area can spring and develop from a host of motivations: the politics of bullying may interest you, its psychology or perhaps you wish to understand your own experience. There are those who counsel, advise or represent. Looking closely at one's motivation, and understanding the roots of one's interest, can only benefit those for whom you work.

The role of aggression

Studies of aggression and bullying have frequently raised the notion of eradicating aggressive instincts, however, aggression should be regarded as part of our potential. Organisations need to take into account the aggressive instincts found within the organisation and to recognise that it is not possible nor appropriate to attempt to legislate against these aggressive drives. Political correctness leads to a world in which spontaneity is absent and fear prevails in human relations. For example, if management cannot exercise legitimate authority for fear that it will be accused of bullying, then we will be developing cultures which are fear-ridden and watchful environments which are explosive and potentially hostile. We will develop the workplace equivalent of our colleagues working as spies, monitoring each other like a police state. The current thinking that tends to focus on the person who is bullied has led to a one-sided appreciation of the area, involving what is difficult to assess. It is easier to describe the impact that the behaviour of others has on one rather than to assess the effect of one's own behaviour on others.

Who are the bullies?

When asked who is capable of becoming a bully the answer must be that the capacity to bully is present in everyone. However the difference between having the impulse to bully and bullying is enormous, just as having murderous thoughts is light years away from actually pulling a trigger. It is the instinct to bully which is present in us all that needs to be understood in all its varieties. At times bullying solely relates to the employees concerned; it belongs to the employees who are directly involved, however the response to workplace bullying is an organisational issue, and the conditions in which individuals are able to bully is also an organisational responsibility. The organisation may encourage through its work practices and structure the base behaviours of men and women to surface.

Personal observations in the field of workplace bullying

Looking at the field as a whole, a paper published in the US many years ago called 'Leave it to George' examined how the difficult patients were left to junior members of staff while the interesting cases were seen by professors. Something similar appears to have taken place with regard to workplace bullying. In the early 1990s the study of the bullying was left to whoever was prepared to take it on – 'George will do it'. Events have moved on with an increasing number of researchers coming into the field, bringing with them a range of experiences. The advances made in identifying and teasing out the subtle elements of bullying are significant. The study of the bully and the bullied requires an assessment of both the conscious and the unconscious factors involved. It is easy to spot a lager lout pushing his or her weight around, however it is much more difficult to spot subtle bullying which creates a huge problem for management.

The needs of those who have been bullied can be great. Those involved in working in this field can testify to the number of unsolicited phone calls they receive, letters that they are sent and the efforts that people will take to seek help. Although the needs of the bullied spring from their wounds, it is often the helper who is left feeling persecuted and overwhelmed. The circumstances involved in bullying can be hugely complicated, of month's or year's standing, some may be genuine cases of bullying, while others are not. What is required is for organisations and not solely individuals to respond to what can be the infinite requests of thousands of individuals who seek guidance, consultation and sometimes solace.

Bullying as a symptom of organisational dysfunction

Specific incidences of bullying require bespoke interventions, tailor-made to the organisation involved. There is also a need for courage, with management or the individual being prepared and able to stand their ground.

Let's look at an example in which bullying was a symptom of the design of the organisation. A chartered engineer who worked in a lighthouse service spoke of being humiliated in front of his staff by the deputy engineer: 'His main method of bullying was to come into my drawing office and take me down in front of my draughtsman.' This is very typical but another observation was even more important. He said there were two bullies in the organisation and commented: 'It is odd that both bullies are deputy heads of departments. The second man is also unqualified having come up through the clerical ranks. I wonder if the bullying commences at this stage as a result of a lack of confidence.'

It is not surprising that bullying was seen to reside in deputies. Deputy heads of anything from government to industry are often in thankless roles, neither having ultimate power nor kudos but often being required to take on the henchmen and lieutenant roles. In this example, the potential to bully resides not just in the individual but also in the role. Not only is firm action required on the deputies' behaviour but also there is a need for an analysis of the deputies' roles in organisations. There are many instances from my practice where aggression has been linked to the role. In another instance an organisation rid itself of the bullying manager to find that the successor also bullied the staff even though they had chosen a totally different personality. In the end, the organisation dispensed with the role. What is required is an analysis of the organisational structure, not merely an investigation of individuals. This approach means that the incidence of bullying can act as a signal that the organisation requires a comprehensive organisational review. An excessive emphasis on action, together with the pressure to deal with bullies, can blind the organisation to two key matters. Policies, statements and procedures drawn up within organisations invariably reflect the character and style of the organisation. An organisation investigating itself faces conflicts of interests between doing justice for the employee by looking into accusations and wishing to keep the bullying quiet with the result that those responsible for making the decisions may form prejudiced views. It is difficult for a senior manager to keep an open mind when distressed employees recall their torments at the hands of a bully. How does one manage justly when the crowd is baying for blood? Organisational procedures need to balance the rights of the alleged bully with the need to end the torment of those bullied. For management, as identified in *Bullying at Work* (Adams and Crawford, 1992), the issue of truth and lies is paramount. But whom should you believe?

The development of policy, procedures and law

The need to know how best to deal with bullying has led to the development of the area of policy and procedures. There is however a limit to the effectiveness of this approach. Policies and procedures alone cannot stop bullying. What is required is firm management and the possibility of punishment, giving the sense that you cannot get away with it. Where an organisation has a policy on bullying this usually means that one or two individuals have taken pains to write it with or without expert advice. Sometimes the policy is borrowed or copied and is inappropriate for the organisation concerned. What is right for a multi-national company may be useless in a family business. What works in a school setting may have little relevance to a chain of bakeries.

Policies and procedures can create the illusion that the organisation has tackled the problem. It is argued that having a policy is a huge step forward, but concern about bullying needs to be lodged in people's minds not in the written word. For many employees bullying at work means nothing. To be effective policies need to be backed by designated groups who are responsible for the sensitive dissemination and maintenance of policy. Yet despite having policies in place problems may arise, for example, strict policies on food hygiene in restaurant kitchens are great in theory, but must be applied in practice. Policies do not stop individuals picking their noses while preparing food for the public.

In an anonymous example one can see the pitfalls that may be encountered. In essence, it is the follow-up procedures that make the difference in dealing with bullying at work.

Case study

A head of a division systematically bullied around 20 staff, many of whom left, over a period of years. A new deputy head of the organisation, on hearing first hand about the events, decided to take the matter seriously. My role was two-fold. I arranged confidential sessions for the staff who were bullied, the alleged bully and others in the department. A limited number of consultation sessions were made available. In addition the senior managers were consulted about how they wanted to deal with the organisational issues that might arise.

Several features in this scenario stand out. A deputy director and senior member of personnel wanted the bullying to be investigated. However the other members of the organisation refused to believe bullying existed and the director and personnel professional were isolated. Paradoxically the others wanted to distance themselves from any possibility of being damaged by any flak or fallout that may occur. To the employees in the division, the new deputy director offered some hope that the tyranny might end. No senior manager until that time had taken their concerns seriously. Among the bully's traits was to single out for criticism individuals who were having difficulties in their lives, such as a miscarriage, bereavement, an ill child or a house move. As a consequence nobody dared tell the bully of any personal difficulties and they felt relieved when someone else in the team was singled out for systematic persecution. To the bully, the deputy director became an ogre whom she accused of being a bully himself. This situation began to be an embarrassment to the organisation. What if the story got into the press? The organisation feared publicity, not because of damage it might do to their image, but rather because it could open up a flood of more employee complaints of bullying.

With support, the personnel manager and deputy director stood firm while statements were taken and a formal investigation was undertaken. During this process, the alleged bully displayed such deviousness that he managed to undermine further his credibility and his categorical denials of allegations of bullying.

The bully contacted potential witnesses and discussed colleagues with outsiders, contrary to the rules of strict confidentiality for all the formal proceedings. This bully also made unfounded allegations about the deputy director to the director, however the director continued to support his deputy. The bully then went off work sick and the organisation began to receive sick notes from the bully's doctor suggesting that he was suffering from stress and depression and alleging that his ill health occurred as the direct consequence of the investigation. During this time the bully continued to attend conferences, which were also attended by those he had bullied. The organisation recognised that this bully should never be allowed to return to the workplace.

What was the evidence that this employee was a bully? Two pieces of evidence support this conclusion. Every employee who came to the consultation sessions, male and female without exception, broke down and cried at their first meeting. None of these employees had been sent to the sessions; they all had come of their own volition. All of these people were well-balanced, bright individuals with no previous history of being bullied. Each had gained a reputation in their specialisms and within the organisation.

Supporting the supporters and investigators

The individuals who investigate cases of bullying require their own support and backing. Frequently investigators report that they are bullied and are accused of being bullies during the investigations. These individuals must be given the support they require in order to contend with the complex and difficult situations they find which may include being obstructed by the organisation fearful that an investigation will turn up issues it would like to keep under wraps.

Many employees on the edges of bullying scenarios feel that their role is rarely acknowledged. The friends and colleagues may witness the bullying, sometimes having to pick up the pieces of their crying, distraught peer. When the bullying goes on for months, this role can take its toll with the result that their capacity to work decreases and they become more prone to ill health. Friends and colleagues face a dilemma: if as witnesses they become embroiled in the fray, they may risk their own position within the organisation; if they don't get involved, they may feel they stood by and did nothing, the organisational equivalent of watching a mugging on a daily basis. The role of colleagues is important as bullying often takes place in private; but the factors that scare a bully most are the possibility of more than one person getting together to complain and the increased likelihood of their behaviour becoming public.

Union officials, personnel departments' management and directors are required to manage organisational paradoxes. They try to balance the needs of the organisation on the one hand with the welfare of the employees on the other. Taking the side of the bullied person brings about the risk of being accused of bias, as you are not taking account of the whole picture. If you tell the bullied employee that they should speak to the bully, you may be seen as taking a reasonable approach to an unreasonable issue. Meanwhile the investigators will be asked about where their loyalties lie. Throughout the whole process there is a strong possibility that those undertaking the investigation will be criticised for acting too quickly or too slowly or for failing to keep everyone informed about what is going on. It is not surprising therefore that management and personnel dislike dealing with these cases. Those who have experience in this type of work are aware that it is impossible to get it right for everyone. While one can struggle to understand the different forces at play and try to do what is best, it will be rare that you will ever be thanked for your efforts.

The bullied as bully

Persistent aggression requires a tough response even though this may be fraught with difficulties. The risk of not taking a firm stance is that the bully may suffer bullying from their accusers. A well known union, having developed a policy on bullying at work, described the problems that it had encountered in its implementation. The union said 'despite a health warning in the policy asking members to take care before leaping to the conclusion that they had been bullied many of the members were determined to see themselves as victims of bullying even when it was them that were doing the bullying'. In another situation a woman came to me about being bullied where she lived. She said: 'since having taken the resolve to stand firm, matters have escalated and I have two main fears.

1 That in the fighting, I am myself becoming a bully, a difficult and unacceptable person.
2 I have positive and stalwart support from some of the other tenants but fear that any contact with them may result in having their lives wrecked also.'

In Novell's study of bullying by e-mail, which was described as flame-mail (Novell, 1997), the most common reaction (31 per cent) was to reply with another flame-mail. Aggression begets further aggression. Notions of winning and losing and giving people a dose of their own medicine come into play.

In Peter Randall's comprehensive book, *Adult Bullying* (Randall, 1997), among the many features identified is the fear, that people being bullied experience, that they might lose control if they confront their persecutor. Hitting someone at work may feel right but is potentially disastrous. What is hopeful is the increasing recognition that hidden psychological factors operating in workgroups do not yield to conventional management or disciplinary structures. For those people who view human relations issues as soft issues, how is it that they are so poorly understood and even less well managed? If they're so soft, then any idiot should be able to do it and any fool should be able to spot bullying. But it's not the case. If you and I want to disguise what we are up to we can.

Training

If people are encouraged to discuss their experiences and open up on a one-day training event, then it is vital that the facilitator has the essential skills to help the participants if their distress comes to the surface. Some training delivered to address bullying at work creates the atmosphere and connotations of dog training and behaviour modification. The approach employs a subtly authoritarian approach with the experts having all the answers at their fingertips without the ability or awareness to understand the depth and complexity of the situation. The proliferation of experts who, like salesmen, offer easy solutions and come with fixed agendas is worrying. These rote-learning approaches are eagerly grabbed by organisations that wish to care and are hungry for insight into the field. However most bullying is far from simple and obvious and doesn't fall into neat categories. Any approach to dealing with bullying will involve a host of interwoven areas including:

- the management and behaviour of groups at work;
- human development and the part psychology plays in bullying;
- the varieties of bullying and the different responses they require;
- the notion that bullying takes place within a wider system in which the problems of the bully may feed into the agendas of the bullied;
- legal, union and industrial relations.

By drawing attention to the possibilities and pitfalls of designing training, policies and procedures, which are parts of an organisational intervention it is important to reiterate the fact that brave and good intentions do not necessarily result in the intended consequences. The following example illustrates this problem.

The problems of the head-on approach

Case study

A very large organisation (43,000 employees) had substantial problems of bullying. Initially, there was an invitation to run some pilot studies. Two voluntary workshops on bullying were undertaken and on the surface they did not appear to go well. The discussions on bullying were rather academic and stilted. Unknown to the consultants certain individuals attending the workshops had been nominated by their superiors who thought it would be good for them. The resentment of these nominated participants at having to attend a workshop resulted in anger with their nominees. This anger was directed at the consultants.

At the end of a very heavy day, the three consultants involved in mounting the event retired battered and bruised, holding sharply differing views. It was the host organisation that had proposed the workshops. The mistake was to accept its model for approaching bullying. Some participants were unforgiving and critical of the workshop method. One of the consultants wearily said he thought we should not continue with the workshops. The alternative view was that in the small groups we had run, the prescience of different levels of management in the groups had led to some critical exchanges. It was important that the dust was allowed to settle and that time was taken to think things through.

The whole event was then completely redesigned. First, bullying had been tackled head on and this had been counter-productive. The best approach in this work is to use a more side-on approach. Furthermore one day was absurd; too short to get into anything valuable. As a result a working conference with the title 'Aggression, power and hierarchy' was designed. This was a three-day working conference which was attended by all levels of the organisation. The conference was advertised throughout the organisation and only volunteers could come. The pre-conference literature described the aims and methods in detail. One of the designated objectives was 'to explore what assumptions underlie the basic rules for the management of power, control and influence'. The conference was also staffed with consultants from within the organisation as well as from the Tavistock Clinic, London. This balance of having insiders and outsiders working together proved very important for an organisation such as this where it was not easy to accept outsiders. The outcome was very interesting. There were five times the number of applications we could deal with and we had to mount further events.

These bespoke interventions have been rigorously assessed for their suitability and value to the organisation. The quality assurance unit of the organisation wrote a 100-page report on the work undertaken and backed it wholeheartedly. What was learned from this experience? The conference allowed examples of bullying, misuse of power, hierarchical relations to emerge in a way that all who were present could see. This was only possible because the environment that was created was safe enough to allow the participants to take risks. It was shown that hitting the organisation head-on did not achieve the objective. It is interesting to note that Littlewoods (1998) and the Manufacturing and Finance (MSF) Union sponsored a bill under the umbrella of *dignity* at work, not bullying at work.

The Dignity at Work Bill (1966)

The initiative by Chris Ball of the MSF Union and Lord Monkswell and those around them to draft and press the case of bullying in the House of Lords threw up many questions. The informed standard of the debate was excellent, as was the level of briefing of those taking part. Let's consider Lord McCarthy's response to the Dignity at Work Bill (1966) in which he clearly stated how an individual who is bullied could leave an organisation and claim constructive dismissal. Lord McCarthy went on to discuss the weaknesses of this course of action: 'First the employee must resign and take the employer to an industrial tribunal. If the employee loses, he or she will not get the job back. They are out. The onus is on the employee to prove that 'the employer's act has destroyed fundamentally the possibility of continued employment. I suggest, he said, that a very great deal of bullying can take place before one reaches the point at which a court will say that there has been a fundamental breach of the employment contract.' The difficulty here is that these processes take a considerable length of time when the employee concerned is least able to fight or to pay the high financial and psychological cost.

Lord McCarthy drew attention to the Criminal Justice and Public Order Act (1994) which makes it an offence to cause 'intentional harassment, alarm or distress' which, if proven, might result in an employer being convicted. He concluded 'that is not what we need or what we want. We do not want revenge; we want reconciliation and acceptance of responsibility. We want, as Lord Haskel said, a higher level of industrial relations, and in the end we need a civil remedy'.

Reconciliation following instances of bullying, while laudable, is unrealistic in many cases. Bullying involves aggression that has got out of hand and the hurt can continue long after the event. Like any trauma, bullying requires to be worked on over a long period of time if it is to be resolved. Is bullying a crime? At times the language used would suggest so: perpetrators, victims, investigations, targets, and so on. Edward Glover in his book *The Roots of Crime* (Glover, 1960) said that crime was part of the price of domestication of a naturally wild animal. In this respect, bullying reflects unsuccessful domestication in the human race.

Freud's discussion on whether civilisation has matured over time and has relinquished primitive instinctual wishes is important. Freud argues the case in 'The future of an illusion' (Freud, 1928) that 'cultural developments lie ahead of us in which the satisfaction of yet other wishes, which are entirely permissible today, will appear just as unacceptable as cannibalism does now'. Freud goes on to say: 'There are countless civilised people who would shrink from murder or incest but who do not deny themselves the satisfactions of their avarice and their aggressive urges and who do not hesitate to injure other people by lies, fraud and calumny, so long as they remain unpunished for it.'

What we know is that increasing the awareness of workplace bullying has helped many people identify what is happening to them: it is a diagnostic tool. This field will continue to extend into other areas. It has already, for example, we see bullying in the community, bullying by the public of shop assistants or GPs' receptionists, bullying on the roads (road rage), bullying from the anonymity of the telephone, and trades which sanction bullying by their aggressive selling. We also see bullying by religious communities of those who stray, bullying of patients in hospitals and the elderly in homes and intergenerational bullying at work. The debt collector can be seen as a bully, as can the policeman. The beggar can be seen as a bully and so can the multinational organisation. Rules, regulations and laws can be used to bully – what is a perfectly acceptable sausage in Britain may be banned in Brussels.

Note

Thanks are due to Dr Guy Michell for his invaluable contributions in the preparation of this chapter.

References

Adams, A. and Crawford, N. (1992) *Bullying at Work*, London: Virago.
Crawford, N. (1987) 'On groups', paper no. 63, London: Tavistock Clinic.
Crawford, N. (1991) 'The psychology of waiting and reception in general practice', paper no. 102, London: Tavistock Clinic.
Crawford, N. (1996) 'Primitive life at work', paper no. 168, London: Tavistock Clinic.
Crawford, N. (1997) 'Bullying at work: a psychoanalytic perspective' in J. Orford, and J. Beasley (eds) 'Bullying in adult life', special issue, *Journal of Community and Applied Social Psychology*, 219–25.
Dignity at Work Bill (3rd reading), House of Lords, 4 December 1996, Hansard.
Einarsen, S. (1999) 'The nature and causes of bullying at work', *International Journal of Manpower*, 20, 1 and 2.
Freud, S. (1928) 'The future of an illusion', *Collected Works*, Hogarth Press, vol. 21, 5–56.
Glover, E. (1960) *The Roots of Crime*, London: Imago.
Littlewoods (1998) *Promoting Employees' Dignity at Work*, Liverpool: Littlewoods.
Novell UK (1997) *Shaming, Blaming and Flaming: Corporate Miscommunication in the Digital Age*, Bracknell: Novell.
Novell UK (1998) *Spanner in the Works*, Bracknell: Novell.
Randall, P. (1997) *Adult Bullying*, London: Routledge.
Rayner, C. (1999) 'From research to implementation: finding leverage for prevention', *International Journal of Manpower*, 20, 1 and 2, 28–38.

CHAPTER THREE

Trauma, duress and stress

MICHAEL J. SCOTT AND STEPHEN G. STRADLING

Introduction

Managing bullying at work involves not only identifying incidents but also dealing with their consequences. One of the most severe consequences that may ensue for a victim of prolonged and unremitting bullying at work is that they develop symptoms of post-traumatic stress disorder (PTSD).

What is post-traumatic stress disorder? PTSD was first included among the Axis I anxiety disorders in the third edition of the *Diagnostic and Statistical Manual of Mental Disorders* (DSM-III) in 1980 (American Psychiatric Association, 1980). The current diagnostic criteria, as set out in the fourth edition (DSM-IV) (American Psychiatric Association, 1995), are that the person must:

(a) have witnessed or experienced a serious threat to life or physical well-being;
(b) re-experience the event in some way (e.g. through nightmares and 'flashbacks');
(c) persistently avoid stimuli associated with the trauma or experience a numbing of general responsiveness;
(d) experience persistent symptoms of disordered arousal (usually manifesting as irritability and hypervigilance);
(e) suffer symptoms for at least a month.

PTSD is unusual in including aetiology as part of the definition: 'its definition explicitly assumes that a particular kind of event is central to the pathogenesis of the disorder' (Duckworth, 1987: 176). The presumption, supported by the syntax of the definition – 'the event' or 'the trauma' – is thus of a single, acute, dramatic episode. Criterion (a) makes plain that just witnessing the event can trigger the disorder – persons on both sides of the fence at the Leppings Lane End at the Hillsborough football ground disaster in 1989, both spectators and emergency services personnel, developed PTSD symptoms.

The PTSD sufferer experiences three groups of symptoms and diagnostic instruments for the detection of PTSD typically embody this presumption of a single episode. For example (from Carlier et al., 1998):

Re-experiencing the event:

- 'I thought about the event regularly, even if I didn't want to'
- 'Sometimes images of the event shot through my mind'
- 'I had the feeling I was reliving the event'
- 'I felt very bad, or got upset whenever I was reminded of the event'

Avoidance of the event and its representations:

- 'I did my best or forced myself not to think about the event'
- 'I have been avoiding people or things that remind me of the event'
- 'I had the feeling that the event was a bad dream, as if it did not really happen'

Hyperactivation since the event:

- 'I have been more nervous and more jumpy, for instance if I hear an unexpected sound'
- 'I have had trouble concentrating'
- 'I have been more apt to be impatient or lose my temper'

Individuals may be traumatised by a variety of intrusive, overwhelming events, e.g. natural disasters, rape and other criminal acts. Some of these may occur in conjunction with their employment, e.g. trench warfare, being involved in or having to deal with the aftermath of a road traffic accident and assault at work. Even so, a single instance of bullying at work – while unpleasant, unwelcome, intrusive and unsettling – is surely unlikely to be of sufficient magnitude to trigger PTSD.

Some recent research, however, has suggested that PTSD symptoms of intrusive imagery, avoidance behaviour and disordered arousal – criteria (b), (c) and (d) – may develop as a result of exposure to an unremitting sequence of workplace stressors – such as a prolonged episode of workplace bullying.

This phenomenon was first noticed in the USA. Ravin and Boal (1989) reported six cases of PTSD symptoms arising from employment-related stressors. In five of the six there was no clear 'single, overwhelming catastrophic experience' (p. 6) instead 'the precipitating stressor-event is but the last of a series of events'. For example: A. N., a male supervisor, 'was experiencing recurrent and intrusive recollections of the reprimands he had received, especially the last' (p. 11), 'not one of which probably would have met the criteria of *Diagnostic and Statistical Manual of Mental Disorders*, third edition, revised (DSM-III-R) for the kind of event necessary to initiate the syndrome' (p. 12). All five cases were experiencing (multiple) intrusive images, avoidance and disordered arousal.

A number of similar cases of PTSD symptoms arising from employment-related stressors have been treated by the first author and two are outlined here.

Case study 1

Charles, a middle manager in a large organisation, was subject to 18 months of persistent 'dumping' of work on him by a superior driven by the anticipation of a merger with a rival organisation. His boss never said 'no' to his superiors, but delegated all the extra work to Charles and his staff. This is not outside the range of usual human experience and so does not in and of itself meet the stressor criterion (a) of the *Diagnostic and Statistical Manual of Mental Disorders* (DSM). Charles prided himself on his fairness as a manager, but was denied the resources for his staff to do an adequate job. He became nauseous at the sight of his work and avoided it. He had flashbacks to a number of encounters with his boss over the previous 18 months. At home he was uncharacteristically irritable with his family. When assessed after being off work for six weeks, he had all the hallmarks of PTSD – save the trauma – and was also severely depressed.

Case study 2

Disa, a woman police constable (WPC) for 12 years, had deliberately sought out many of the most challenging assignments in police work. She had worked in the Criminal Investigation Department (CID) and on the drug squad and had volunteered for the child abuse team. The proximal stressors Disa experienced prior to developing PTSD symptoms were, objectively, no different to those she regularly encountered in her work and was not outside the range of the usual experience of any conscientious and ambitious WPC. She had been off work for six months when referred. She reported that no single event triggered the PTSD symptoms but that it was the cumulative effect of her encounters that came to debilitate her.

Disa described four stressors she had endured and which were now figuring in flashbacks. In one episode, for example, she was rushing a young mother to hospital with her baby only for them to find on arrival that the baby was already dead. The two of them just sat on a bench outside the hospital holding the baby for half an hour. Another involved attending a post-mortem on a murder victim who was the same age as Disa. After each event she had been able to get on with her work, but eventually, as she put it, 'the sand just filled the bucket until it overflowed'. She was showing considerable avoidance behaviour, making excuses not to return calls to colleagues or to friends who were police officers, and was worried that she was driving her boyfriend away with her 'temper tantrums'. She was also severely anxious and depressed.

The manner in which these two cases meet all but the stressor criterion (a) of DSM-IV is shown in Table 3.1. Axis IV of DSM-IV invites characterisation of the severity of psychosocial stressors experienced by an individual, distinguishing acute stressors from enduring circumstances. It would now seem that either type may lead to intrusion, avoidance and disorder of arousal whether

Table 3.1 How cases meet DSM-IV criteria A to E (from Scott and Stradling, 1994)

	Case 1	Case 2
(a) Experienced an event outside the usual human experience	No	No
(b) Traumatic event is persistently re-experienced (at least one symptom)	Yes Recurrent and intrusive distressing recollections	Yes As Case 1 plus recurrent distressing dreams of event
(c) Persistent avoidance of stimuli associated with trauma or numbing of general responsiveness (at least three symptoms)	Yes Efforts to avoid thoughts or feelings associated with the trauma Efforts to avoid activities or situations that increase recollection Markedly diminished interest in significant activities	Yes As Case 1 plus sense of foreshortened future
(d) Persistent symptoms of arousal (at least two symptoms)	Yes Difficulty falling or staying asleep Irritability or outbursts of anger Difficulty in concentrating Physiological reactivity on exposure to events that symbolise or resemble trauma	Yes As Case 1 plus hypervigilance, startle response
(e) Duration of symptoms last at least one month	Yes	Yes

through a single overwhelming experience of great intensity but of a (relatively) short duration (currently the typical case – PTSD) or through prolonged duress brought about by a series of unremitting though individually (relatively) less intense circumstances. We call this prolonged duress stress disorder (PDSD).

The following case study shows how someone developed PDSD through bullying at work.

Case study 3

Barry was referred for counselling by his GP as suffering from depression. Barry had been made redundant four years previously after only three months with his employer. Barry clearly missed working and always looked at the job advertisements in his local paper but said that he had lost confidence in his ability to work. Enquiry about the onset of his lowered mood revealed that he was already very low when he took the job from which he had been made redundant. He had taken up this position after three months of unemployment having left his previous job because he 'couldn't take any more'. When asked details of this previous post, in a major retailer, he said 'I can't talk about it, it's too upsetting, I had been with them 15 years'.

The counsellor made a note of this comment and began a standard cognitive-behavioural programme for depression (see Scott and Stradling, 1998) with an initial emphasis on scheduling in activities that were potentially uplifting. Barry decided that he would try to spend some time each day in his garage making doll's houses, a hobby that he had given up about five years ago. He said 'anyway it gives me an excuse to get away from the family'.

Reviewing Barry's homework assignment at the next counselling session revealed that on some days he had been able to make doll's houses and on others not. In the light of his comment at the initial assessment, the counsellor asked 'on some days do you get so distracted by a memory that you can't do anything?' Barry became tearful and began to describe the 18 months of bullying at the retailers.

He said that everything he did was criticised by his boss. The last straw appeared to have been when a report he had produced was flung aside by his boss and the latter swore at him. His demoralisation was complete when the senior managers took no action against his boss despite the latter having caused serious problems with other colleagues.

The counsellor discovered that Barry:

- experienced disturbing intrusive imagery of the various conflicts with his boss;
- showed avoidance behaviour, switching the TV off when the retailer's products were advertised, avoiding even those former colleagues who were supportive and isolating himself from all other possible sources of support;
- had become uncharacteristically irritable and was regularly experiencing sleep disturbance.

In addition, Barry was embarrassed that he should have had such a reaction to 'mere' bullying and, as a consequence, he had not been forthcoming with professionals about the real source of his problems.

Barry met the symptom criteria (b), (c), (d) and (e) for PTSD and was suffering PDSD brought about by bullying at work. The matter finally came to light four and a half years after leaving his employment as a result of workplace bullying. The counsellor commenced a programme of cognitive-behavioural treatment (Scott and Stradling, 2000) to alleviate his distress.

Treatment

Clinical experience to date with PTSD and PDSD suggests that aetiology has consequences for treatment. Treatment for PTSD typically involves:

- desensitisation to intrusive imagery by controlled re-experiencing of the trauma via, for example, daily playing at a pre-set time of a 15-minute audiotape prepared by the client which describes the trauma;
- tackling avoidance behaviour through graded exposure to feared situations;
- anger control training for the irritability which disordered arousal commonly manifests;
- treatment as appropriate for any co-morbid disorder (depression is common).

PDSD patients are more resistant than PTSD cases to making and listening to trauma audiotapes (though of course a significant minority of the latter also show resistance – Scott and Stradling, 1997). The more diffuse nature of the PDSD sufferer's experience means that they tend to anticipate little positive benefit from being able to 'digest' an upsetting episode – their outcome expectations are negative. They also believe that so much has happened to them that focusing on one episode will trigger memories of others that they would be unable to tolerate. While desensitisation is preferred for chronic PTSD cases, we have suggested that for PDSD the emphasis should be on cognitive restructuring and assistance with 'balancing out' negative memories with positive ones (see Scott and Stradling, 1992).

In his recent seminal text on the treatment of PTSD, Meichenbaum (1996) identifies seven classes of PTSD clients, of which those suffering PDSD is one. How common is PDSD? As yet we do not know, but providers of workplace counselling and emotional support, both in-house and via employee assistance programmes, may expect to encounter occasional cases of PDSD.

Bullying at work as a stressor

Of course bullying is just one of a number of sources of stress at work. Table 3.2 gives a comprehensive list of job stressors.

Counselling victims of bullying and harassment

A general cognitive-behavioural model of stress – the balance model of stress (Stradling and Thompson, 1997; Scott and Stradling, 1998) – may assist in the counselling of victims of bullying and harassment at work.

Psychologically, when perceived demands are out of balance with perceived resources, an individual is stressed. The balance model of stress (see Figure 3.1) represents and develops this formulation. When demands and resources are appraised by the person as in balance, the individual is in the stress-free zone. An excess of demands leads to overload and substantial under-demand leads to underload. The moveable fulcrum represents individual differences in interpretations of situations where objectively equivalent circumstances are interpreted as 'a stimulating challenge' by one person and as 'overwhelming' by another.

Table 3.2 Job stressors

- Is your workplace safe (non-toxic) and comfortable (ergonomic)?
- Is your workplace free of bullying and discrimination (harassment)?
- Do you have variety in your job (variety)?
- Do you feel you are fairly rewarded for the work you do (equity)?
- Do you have enough time and resources to do your job (workload, workpace)?
- Are you clear exactly what is expected of you in your job (role clarity)?
- Is your job free of conflicting demands from others (role ambiguities, boundaries)?
- Do you have to do things you wish you didn't have to do (value-matching)?
- Are you free to make decisions about the way you do your job (autonomy, discretion)?
- Are you involved in or informed about important decisions at work (participation)?
- Do you have responsibility for other people or responsibilities that affect the welfare of others (responsibility)?
- Is your job under threat or your work being reorganised (job security)?
- Do you have the opportunity to change the job you do at work (internal labour market)?
- Could you get the same or a better job elsewhere (external labour market)?
- Are you clear how your work contributes to the success of the organisation (recognition)?
- Do you have a supportive supervisor (interpersonal – vertical)?
- Could you confide in the people you work closely with (interpersonal – horizontal)?
- When work gets tough do you get plenty of support at home (home/work interface)?

Figure 3.1 Balance model of stress

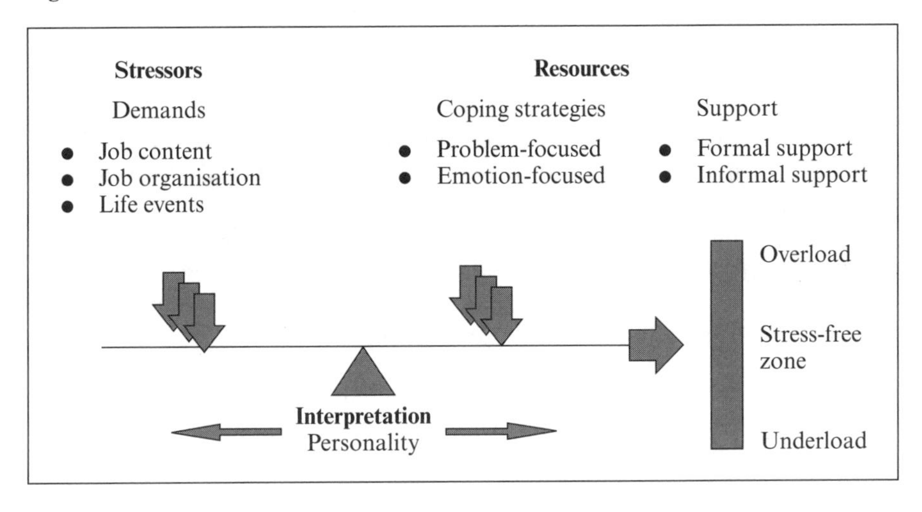

This is in line with Epictetus' dictum, from the 1st century AD, that 'people are disturbed not so much by events as by the views which they take of them' which has informed the development of modern cognitive-behavioural approaches to therapy (Scott et al., 1995).

As noted in Figure 3.1, bullying and harassment at work may load on the 'demands' side of an individual's equation. In general, distinction must be made between remediable and non-remediable aspects of work stressors. Bullying is remediable.

The 'resources' are divided into two broad types:

- coping strategies, both problem-focused and emotion-focused;
- support, both formal (provided by the organisation) and informal (provided in an *ad hoc* manner by colleagues, family and friends).

This formulation allows clear identification of which aspects of the situation management should address in:

- formulating an organisational stress strategy as part of a proactive human resource policy (*not* just as part of a reactive staff welfare programme – stressed human resources cost money even before employees go off sick);
- dealing with workplace stress.

Individual repair, on the other hand, whether administered by the well-meaning or, in the case of the severely distressed, by professionals, recognises that it is typically unable to exercise control over workplace demands and seeks to enhance individual resources. This will involve facilitating problem-focused strategies to re-establish the individual's control over the situation and emotion-focused strategies to re-establish control over feelings. Counselling to this end may be successfully undertaken in a small group, as well as an individual, setting (Scott and Stradling, 1998).

Individual repair changes the person – in extreme cases it may need radically to reshape their interpretation of the situation. Organisational repair changes the conditions under which persons operate.

Successful individual repair will restore the balance between perceived demands and perceived resources but the sought-for outcome is *not* the removal of demands (this would create stressful underload), but establishing a balance between demands and resources that is seen as manageable. This may mean:

- changing the interpretation of the original demands,
- replacing them in whole or part with lesser demands, or
- amplifying the resources available to the individual.

The successful organisation will take responsibility not only for repair but also for reform. It will remedy current sources of distress, develop adequate monitoring mechanisms to prevent re-occurrence and, having alleviated distress, will endeavour to go on to improve the job satisfaction of its doubtless currently over stretched workforce.

And Barry? He is now much improved as a result of the counselling repair work – but he'll never be the same person he was before being bullied at work.

References

American Psychiatric Association (1980) *Diagnostic and Statistical Manual of Mental Disorders*, 3rd edition, Washington, DC: American Psychiatric Association.

American Psychiatric Association (1995) *Diagnostic and Statistical Manual of Mental Disorders*, 4th edition, Washington, DC: American Psychiatric Association.

Carlier, I. V. E., Lamberts, R. D., Van Uchelen, A. J. and Gersons, B. P. R. (1998) 'Clinical utility of a brief diagnostic test for post-traumatic stress disorder', *Psychosomatic Medicine*, 60, 42–7.

Duckworth, D. H. (1987) 'Post-traumatic stress disorder', *Stress Medicine*, 3, 175–83.

Meichenbaum, D. (1996) *Treating Post-traumatic Stress Disorder: A Handbook and Practice Manual for Therapy*, Chichester, Wiley.

Ravin, J. and Boal, C. K. (1989) 'Post-traumatic stress disorder in the work setting: psychic injury, medical diagnosis, treatment and litigation', *American Journal of Forensic Psychiatry*, 10, 5–23.

Scott, M. J. and Stradling, S. G. (1992) *Counselling for Post-traumatic Stress Disorder*, London: Sage (2nd edition, 2000).

Scott, M. J. and Stradling, S. G. (1997) 'Client compliance with exposure treatments for posttraumatic stress disorder', *Journal of Traumatic Stress*, 10, 523–6.

Scott, M. J. and Stradling, S. G. (1998) *Brief Group Counselling: Integrating Individual and Group Cognitive-Behavioural Approaches*, Chichester: John Wiley.

Scott, M. J., Stradling, S. G. and Dryden, W. (1995) *Developing Cognitive Behavioural Counselling*, London: Sage.

Stradling, S. G. and Thompson, N. (1997) 'Stress and distress at work: whose responsibility?', British Psychological Society annual conference, Edinburgh, 2–6 April 1997.

CHAPTER FOUR

Victim to survivor

NOREEN TEHRANI

Introduction

The experience of being bullied can be very familiar, raising painful childhood memories of school or family. But for some people, being bullied is something that only happens in later life with the result that the victim feels totally unprepared and perplexed by what is happening to them. This chapter presents the stories of four people. Each story is used to illustrate the nature of bullying, the effects of the bullying on the victim and how victims of bullying come to understand what had happened. This chapter provides validatory narrative documents that describe the experience of victims of bullying and enables the reader, as witness, to identify with and to learn from the experiences of the bullied. While it is not possible to look at therapeutic approaches in depth, some of the underlying psychological principles and theories are described.

In Chapter 3, the evidence that bullying can result in post-traumatic stress was discussed. The work of Leymann and Gustafsson (1996) and Scott and Stradling (1994) has resulted in the adoption of trauma counselling and debriefing techniques with victims of bullying (Scott and Stradling, 1992). In recent years the growing interest in narrative therapy (Payne, 2000) has led to the use of these techniques with victims of bullying. There are a number of similarities between the structure of psychological debriefing (Tehrani and Westlake, 1994) and the structure of the narrative or life story (McAdams, 1985). However, debriefing is used to help people to focus on individual traumatic incidents while narrative therapy helps to place events within the greater framework of the person's life story. The telling and re-telling of stories as a way of handling distressing events has a long history and is found in many different cultures. It is an essential aspect of the ways of life of the American Indians (Wilson, 1989), Australian Aborigines (Wingard, 1998) in addition to many other cultures (McLeod, 1996). Today, story telling is being used to assist in the healing process for people facing a wide range of personal and psychological problems (White, 1998).

The following stories were collected to show how a number of people experienced bullying. All the stories are based on a real event, however some of the details have been changed to protect the identity of the participants. Using tools from debriefing and narrative therapy each participant was asked to take a single example of the bullying, recreating the words used by themselves and the perpetrator. To the verbal account was then added the recalled thoughts, emotional reactions and physiological responses. The intention of gathering the stories was to provide a testimony, however each of the participants found that, by telling their story in this safe and supported way, it had helped in creating a meaning for what had happened and provided a release from their harmful memories.

Charles' story

Charles was in his early fifties and had been with an organisation as a manager for more than 25 years. Charles enjoyed his role and was happy in the knowledge that he had an effective and contented team. The team was proud of its reputation, its standards of customer service and that it consistently met its budgetary targets. Charles was an active member of his department's management team taking an active part in training and internal consultancy. When the head of department post became vacant, Charles decided to apply for the post as he believed that he had a lot to offer and would welcome the challenge. However, being the only internal applicant he was not surprised to discover he had not been selected.

Kate, an external candidate was instead recruited as the head of department. She was younger than Charles and had gained a reputation for cost reduction in a commercial company. Soon after Kate's arrival her relationship with Charles and the management team deteriorated and the bullying began. Charles was not the only manager to be bullied by Kate as nearly all of the managers had been targets for Kate's bullying at one time or another. However, Charles felt that he was singled out for bullying more than the other managers and thought that this may be due to his unsuccessful application or because he was the most senior member of the management team. Charles tells two stories:

Team building

The first time it became obvious that something was going wrong was at a meeting of the whole department that was to be a team-building exercise. Kate had asked Bob to call the meeting and said that the meeting would provide an opportunity for the managers to enthuse their staff. The managers had high hopes for the meeting.

Kate arrived at the meeting 10 minutes late. She immediately left the room to organise the coffee. Charles thought that this was very strange and could not understand why she was not taking charge of the meeting. On her return to the meeting room Kate sat down.

Kate: OK then chaps how is it going?
Bob: We were waiting for you so that we could start.
Kate: Well I'm here now so off you go.
Bob: It's your meeting so the floor is yours.

Kate looked irritated, she looked at Bob and spoke in a raised voice.

Kate: When I asked you to organise the meeting, I expected that you would have planned the agenda.

Charles thought 'what is going on? It is not fair for her to expect Bob to do something without being asked and she is not behaving professionally by taking that tone with Bob in front of his staff'.

Kate: Well if you can't do it then I will have to do it myself. The first thing that I want to say is that it is unacceptable for people to arrive late at my meetings. It shows a lack of respect and also a lack of planning of your time. I want everyone to ensure that they attend on time in future.

Charles was feeling uncomfortable. He wanted to respond but felt unable to do so. He thought 'this is getting off on the wrong foot. Why is Kate taking this negative attitude to the staff? It is counter-productive to upset them in this way. Has she no awareness that she was 10 minutes late herself?'

Kate continued in a similar vein for about half an hour. She then called the meeting to a close and asked for the managers to stay behind. Charles felt upset and disappointed. There was little that could be said which would not make the situation worse. Kate's choice of venue to unleash her negative comments prevented any debate or discussion of team issues.

Kate turned to the management team:

Kate: When I ask a manager to do something I expect it to be done. Bob's lack of planning caused that fiasco. How can I manage the department if that is the sort of support that I get? I expect a better standard of performance from my managers. It looks as if I will have to teach you a thing or two about management and responsibility.

The management team was stunned into silence. Charles felt hot and uncomfortable. Kate was undermining and threatening the managers for a situation that had been caused by her failure to make her wishes known. Charles was unable to let the situation continue without offering some support for his colleague.

Charles: My impression was that you had not asked Bob to arrange the content of the team-building event, only to call everyone together for the meeting.

Kate was getting more and more angry. Charles realised that his intervention had turned the focus of Kate's anger on to him but felt that he had no alternative but to speak in support of his colleague.

Kate: I told Bob to call the meeting and to run the event. I know what I said. Are you accusing me of not knowing what I said?
Charles: No, but we all had the same impression so there must be some misunderstanding.

Charles felt shaken by the events of the afternoon. He had seen a different side to Kate and was concerned about the unnecessary shouting, browbeating and the distortion of the truth. Charles knew that by challenging Kate his relationship with her would become more difficult.

Making the budget

Kate's manager, Judy, called a meeting with Kate and her management team. The purpose of the meeting was to examine the financial performance of the department. Charles was not particularly concerned about his team's financial performance as he knew that he had reduced costs and increased profits. In fact Charles took great pride in the fact that his team was always on budget and consistently profitable. During the meeting the following conversation took place:

Judy: What about your area Charles?
Charles: Actually we seem to be performing well on budget and should exceed our profit target at the year-end.
Kate: That's not right. You are heading for a year-end loss of £800,000.

This statement was totally untrue. Charles was stunned and could not understand why Kate would misrepresent the real situation.

Charles: I do not understand, that isn't what my figures show. Where did you get that information? Can I see your figures?
Kate: My information is correct. I will gladly show the figures to you after the meeting.
Judy: Well Charles it looks as if you also have some work to do.

After the meeting Charles asked Kate for the figures, but was met with a blank statement that his figures were wrong and that her figures were right and, yes, she would show him the figures when she was ready but not now. Charles felt devastated by this turn of events, not only did he feel that he was made to look incompetent in front of a senior manager but also that he was unable to refute this allegation unless he was given Kate's information which she was refusing to share. He knew that his figures were from the management accounts and therefore were by definition the correct ones and that this allegation of an £800,000 loss was an invention.

A few months later Kate asked Charles to prepare a proposal to reduce the size of his team and to make sure that he included his own post as one of them. This he did willingly because by this stage he could see no other way out of the bullying situation but to leave the organisation. Charles left two months later. He negotiated a financial package that secured his future. Charles used the opportunity of being unemployed to retrain and he is currently working as a stress manager. He finds this new career very rewarding on a personal level and feels that his experiences bring an extra dimension to his work. Kate had her responsibilities reduced within a couple of months of Charles leaving the organisation.

The moral of the story

Charles has come out of this situation having learned a number of things about himself:

- Despite hardship he has the strength to stand up for what he believes.
- Truth will eventually prevail.
- He is a survivor.

Maria's story

The next story involves Maria who was 19 years old when she started to work in a research department. All the other members of the research team were male and there had been some resentment at her appointment. Most of the other technicians were friendly towards her but as they had worked together for many years she found it difficult to become part of the group. One member of the research team, Jim, was particularly difficult. Maria noticed that he had tremendous power over the other team members. Although he was not the most senior of the technicians he was able to get whatever he wanted even when it required the approval of the research manager. Maria could not understand why Jim was able to get away with his irrational and unreasonable behaviour.

The boiling beads

As Maria began to increase her knowledge and use her skills, Jim began to notice her and seemed to show an interest in what she was doing. One day Jim came over to her working area although it was at the other end of the laboratory to where he normally worked. Jim looked irritated. He stood at the end of the workbench and picked up a glass container holding several hundred glass beads which were used to put in test tubes to stop liquids boiling over.

Jim: You think that you are getting on well don't you? Well, I can make you do anything I want you to do.

Maria was not sure what Jim was talking about. Was this a joke? He looked menacing and Maria began to feel apprehensive.

Maria: What do you mean? Why should I do anything you want me to do?

At this point Jim picked up the container and dropped it on the floor. The container shattered into pieces and hundreds of beads rolled all over the floor. Maria stood there speechless. The crash of the glass container and the sight of the beads all over the floor seemed unreal. She could not believe what she had just seen.

Jim: Now pick them up.
Maria: No, I will not pick them up. You dropped them you can pick them up.

Maria was burning with anger, but her voice was precise and measured. She looked around for the others. Despite the noise of the breaking glass that must have been heard all over the laboratory no one came over to see what was going on. Jim looked right into her eyes.

Jim: You will pick them up or things will get worse for you.

Jim walked away leaving Maria looking at the broken glass and the glass beads all over the floor. She wanted to walk off leaving the mess on the floor. She thought of reporting Jim to the head of research. Mike, one of the other technicians came over to her. He spoke in a very quiet voice.

Mike: One of the things you need to recognise is that he picks on people in turn. If you stand up to him it gets worse. The best thing you can do is pick them up and say nothing.

Maria was upset but felt that she had no alternative. Other technicians came over to her and told her of the things that Jim had done to the others, how he had humiliated young technicians until they could stand it no longer.

Following that event there were many other occasions when Jim taunted Maria. She avoided him whenever she could but this was not always possible. He would make personal comments about her and take pleasure in describing his sexual exploits and asking Maria about her personal life. If Maria did not answer him the taunting would go on and on. He had a good memory and would use any personal information he had acquired to include in his taunts.

Maria noticed that whenever he taunted others, no one ever went to their support for fear of what might happen to them. It was as if there was an understanding that if you allowed yourself to be humiliated then you were let off the hook but if you stood up for yourself then you would be continually taunted, your property would be hidden and your work sabotaged.

The tea room

One day after Jim had made it clear that he did not like women working in the laboratory, Maria went to the tea room and found that all the walls were covered with pornographic pictures. Jim was in the middle of a group of men smirking. Maria walked out of the room and wondered what she could do. She decided that the best thing she could do was to cover the pictures with pictures of her own. The next day she brought in some magazine pictures and put them up. When Maria went into the tearoom later that morning, Jim and the others were sitting there waiting for her. They had taken down Maria's pictures. Maria stood at the door and shouted at the men.

Maria: How dare you do this to me! I did not take down your pictures and you had no right to take down mine. You should be ashamed of yourselves! How would you feel if someone treated your wife or daughters the way that you are treating me?

Maria was very angry but she looked around the room of stunned men and realised that they were the cowards, not her. Maria was not going to put up with this kind of behaviour. She went to the toilets and cried, partly with anger and partly because she was upset. As soon as she was able to compose herself she went to see the most senior manager in the research unit and reported what had happened.

Manager: I have heard what happened to you. Perhaps the best thing would be for you to have your tea with the secretaries and not with the technicians.
Maria: I don't see why I should be stopped from going into the tea room and I don't see why I have to put up with this kind of behaviour.
Manager: I will see that the pictures are taken down.

Maria felt let down by the manager but was determined that she would not be forced out of the tea room by a group of cowardly men. She went back to her work. All of the men apart from Jim came and apologised to her for their behaviour.

Maria's stand against Jim spread around the whole research unit and then over the industrial unit. She had gained a kind of notoriety that provided her with some protection against Jim. However, Jim continued to harass other people. Maria realised that Jim's behaviour was not normal and that because of his power over almost everyone he came in contact with no one was prepared to challenge or discipline him. Maria decided that if the organisation was not prepared to deal with Jim then this was not the kind of organisation she wanted to work in.

The moral of the story

Maria left the organisation, her self-image and self-esteem intact. She had learned that

- Some people are allowed to be bullies by the inaction of others.
- It is impossible to deal rationally with people who are mentally ill.
- No one needs to put up with bullying.
- She is a survivor.

Mike's story

Mike had been in the police force until following an injury, he was unable to continue as a police officer. However Mike had enjoyed working for the police and decided that if he was unable to remain as a serving officer he could train to become a scene of crimes officer (SOCO). Mike funded his own training and completed the course with high marks. Mike then joined a police force where he worked for some time without any problems. When a job became vacant in an area of the country he wanted to live, Mike applied and was appointed as a SOCO. It was not long before Mike became aware of the problems at the station to which he had been assigned. Initially Mike could not understand the antagonism that faced him. Nothing in his previous experience in the police force prepared him for what was about to happen.

The expenses claim

The first attack on Mike took place soon after he arrived at the police station. A detective inspector (DI) who had been particularly difficult and dismissive of Mike accused him of falsifying an expenses claim form. Mike was called into the DI's office where the allegations of fraud were made. The DI's raised voice was heard all over the station and when Mike returned to his own office a colleague reminded him of the records that had been kept which proved that he had not falsified the claim. Mike returned to the DI's office with the evidence. Mike was feeling apprehensive as he went into to the office as the previous meeting had been so acrimonious.

Mike: Here are the records that show I have not made a false claim.

The DI looked very angry and began shouting. Mike found this extremely intimidating. He felt that he had to protest his innocence. He felt that his honour had been called into question and he knew that he had never committed fraud in his entire life.

DI: I am f****** watching you and I will f****** get you.

Mike felt hurt and defenceless. He was upset by this encounter and decided that he would report it to his line manager who was based at divisional headquarters.

Soon after the incident was reported Mike was called into the office of a detective sergeant (DS) from the Criminal Investigation Department (CID). The DS was standing up and looked very threatening.

DS: We know what you are up to and we don't like it. You are a f****** civilian and if you don't like it f*** off.

Mike was not used to being spoken to in this aggressive manner. Throughout the encounter the DS constantly threatened him, pointing his finger at him and shouting. His eyes were bulging with anger.

The atmosphere in the station became tenser with the CID officers making Mike's life even more difficult. Mike requested a meeting with the local superintendent with the view of improving working relationships but, despite some assurances that something would be done, nothing changed. Following this meeting no one spoke to Mike – he was sent to Coventry.

The pensioner

Part of Mike's job involved him photographing victims of crime. A few days after the confrontation with the DS Mike had arranged to photograph a pensioner who had been injured in a robbery. The pensioner had to travel some distance from his home to a police station. On the day of the appointment Mike was told that he had to investigate two burglaries and this gave him no time to cancel the meeting with the pensioner. Mike felt bad because he knew it had been difficult for the pensioner to come to the station. The pensioner agreed to come again on the next day. However the next day the same thing happened, this time it was only an attempted burglary that would not normally have required Mike's presence. Mike was told to cancel the session with the pensioner again that meant that the opportunity to provide photographic evidence of the pensioner's injuries would be lost.

Mike: This is not acceptable. An attempted burglary does not need my presence. I cannot let the pensioner down again. We need to get the photographs of his injuries as evidence as soon as possible.

Despite the strong reasons given by Mike to go to photograph the pensioner's injuries his request was turned down. Mike felt that the officers in CID had it in for him and that as a result an innocent pensioner was suffering.

The relationship between Mike and the officers in the CID grew worse. There was a total lack of co-operation and whenever possible obstacles would be put in his way. On one occasion Mike had to arrange his own transport to a scene of crime when appropriate vehicles were available. One day as Mike drove back to the station he was summoned to a meeting with the DS. Remembering the earlier meeting, Mike purchased a small tape recorder so he could record the meeting. This meeting was extremely threatening and intimidatory. Following this encounter Mike felt distressed and a short while later he had an opportunity to speak to a Detective Chief Inspector (DCI) with responsibility for his post. The DCI said that he did not want this kind of thing happening to any of his men.

The grievance meeting

Following the meeting with the DCI Mike was summoned to a meeting with the station superintendent who was very angry and said 'What the f*** have you done? You have spoken out – it is out of my hands'. Mike felt isolated and alone. He felt that he had no protection and felt totally let down. Shortly after the superintendent was replaced by another officer. Soon after the new superintendent (Sup.) arrived Mike was approached in the corridor. Mike knew of the reputation of this new superintendent and was concerned that things might get worse instead of better.

Sup.: You have made a complaint about the DI, and I am telling you that you have spoken out of turn and the DI is not happy and wants to challenge you.

A grievance meeting was called but Mike's line manager was not invited. However Mike asked his line manager to be present. In a pre-meeting with his manager Mike produced the tape and a transcript of what had happened in the meeting with the DI.

The grievance meeting was one sided. Mike was constantly accused of lying. Mike's line manager then called for a break in the meeting and asked Mike's permission to reveal the presence of the tape recording. The tape provided all the proof necessary to prove that the DI had used bullying tactics and the grievance meeting was closed. Following the meeting the superintendent spoke to Mike:

Sup.: I have advised the DI that his behaviour was inappropriate. He has apologised.

Mike had at least expected some action would be taken against the DI. He felt helpless and believed that there was nothing he could do. The system was against him. Bullying was institutionalised and there was nothing he could do on his own to fight it.

Mike left his role as a SOCO. He had worked hard and in previous SOCO jobs had done well. He was unwilling to accept the bullying behaviours of his CID colleagues and had stood up to them. He recognised that he was not fighting a single bully but that the whole culture was one in which bullying was a way of life and that anyone was fair game to the bullying tactics of the CID gang.

The moral of the story

Mike left the police force, his health and self-image destroyed by the months of bullying. It has taken a long time for him to recover but he has learned that:

- Bullying cultures are difficult to challenge.
- It is important to stand up for what you believe and keep self-respect.
- It is better to be injured for what you believe than to join in with the lies.
- He is a survivor.

Julia's story

The next story concerns Julia, the head of a department in a large organisation. Julia had previously worked in another department in the same organisation and over a number of years had established an international reputation for her work. She had been put forward for the role by her previous manager, Dick. There had been a history of bad feeling between the two departments. However, Julia believed that she had the skills to develop the new department and to forge better relationships between the two departments. Before long conflict arose between her and Dick. Julia thought that her success in the new post was viewed as a challenge to Dick's authority. Dick and Julia reported to the same director and when a new director was appointed Dick was able to persuade him that Julia was the cause of the problems between the departments and should be removed. Although initiated by Dick, it was Cyril, the director, who made false allegations and who carried out the bullying which resulted in Julia leaving the organisation. Cyril had a number of meetings with Julia over a period of time, at each meeting he would endeavour to undermine her by refusing reasonable requests and by increasing her already excessive workload. The following is taken from a meeting between Cyril and Julia. The story gives an indication of how Julia tried to maintain her self-respect against the odds.

The monthly meeting

A regular monthly meeting that had been arranged a month earlier was cancelled at the last minute. Cyril's secretary then arranged another meeting at very short notice. Julia did not know if the new meeting was to take the place of the cancelled monthly meeting or was her appraisal. When she arrived at Cyril's open plan office she found that the meeting had been moved to a private room. Before the meeting began Cyril offered Julia a glass of water. This puzzled Julia.

Cyril sat silently, staring at Julia. Julia had a sense that this was going to be another difficult meeting.

Julia: I don't know the purpose of this meeting. There are some things to be discussed from the normal monthly agenda but we should be discussing my appraisal.

Cyril: We will start with your appraisal.

Julia put her files on the table. She had spent the weekend preparing for the appraisal and had brought supporting documentation with her.

Cyril: This is your appraisal. I have not filled in the performance against team objectives, as you have not sent them to me.

Julia: I had hoped that they would be ready for today, but they are not ready yet.

Julia had asked that the results were sent to her before the meeting, but she had been out of the office on business and it had not been possible to pick up the figures.

Cyril (in a very abrupt tone): Are you blaming this on your team?

Julia: No if there is any blame it is down to me. It is my responsibility.

Julia could not understand why Cyril was being so aggressive towards her.

Cyril: Would you like to read what I have written? Some is good and some not good.

Julia began to read the first page of the appraisal. The first section listed some of the things she had achieved but, without comments, it looked more like a shopping list than an appraisal. The second section talked about Julia's approach. It seemed unreal. Julia could not take it in. The whole section was full of unsupported personal criticisms.

Cyril: Do you accept what is said?

Julia thought 'what am I supposed to say?' The pressure was tremendous, she felt dizzy and sick. 'What is the point of saying anything? He has never listened to anything I say.' Julia felt the full emotion of the situation but was determined not to show her distress. She decided to concentrate hard and look right at him.

Julia: I accept that this is your opinion of me.
Cyril: This is exactly what I mean. You will never accept things you don't want to hear.

Julia sat still, trying to think about how she could respond to this pressure. 'What was the point of challenging the statements in the appraisal document?' He had given no evidence of where the statements had come from. Julia had tried over and over again in the past to try to get Cyril to be clear on what he wanted or meant and he had never done so. Why should this time be any different?

Cyril started to read the comments on the second page of the appraisal.

Julia was unable to take in what was being said. There was a stream of criticisms, as the criticisms were made she used her only defence and began to paraphrase and clarify what Cyril was saying, making comments only when the statements could be independently demonstrated as untrue. Everything she said was rejected. She felt that she was in a nightmare, all reality had vanished. She was trapped.

Cyril picked up the self-appraisal review Julia had been asked to prepare.

Cyril: You sent me this with your feedback on your objectives. What you have done here is to present only the things which suit you.
Julia: The self-assessment review includes information on my objectives and also the information on my performance in my role. I understood that this was what was required for the appraisal.

Julia had tried to get Cyril to recognise her job description required more from her than the objectives. Clearly Cyril had no intention to look at her role and wanted to base the appraisal purely on the objectives. Julia decided that she would challenge this as the appraisal process was new and did require the inclusion of performance against job description.

Julia: An appraisal is not only on annual objectives, it should also be on performance in the role. The appraisal should also take account of the difficulties encountered in the year. My previous manager did not give me objectives on every area of my job description so I thought it was important to give an indication of my achievements in those areas.

Julia was feeling quite cross at this point. Prior to the meeting she had spent time trying to find out how the new appraisal system should work. This appraisal meeting was nothing but a one-sided attack. There was no evidence presented.

Julia felt like walking out but knew that if she did then she would be in a much worse situation.

Cyril: The first objective is OK, but you went on to complete your second objective despite being told not to do so.

This was totally untrue; Julia had been prevented from undertaking work on the second objective by Cyril and had the evidence to prove it. She was not going to leave this unchallenged.

Julia: I stopped the second objective as soon as you asked me to do so.
Cyril: But you would not have stopped if I had not told you to do so.

The whole irony of the situation was now clear. Initially Cyril had claimed that Julia had not done as she had been told. Cyril, having had to agree that Julia had stopped working on her second objective, now was able to claim that she had failed on the objective.

Cyril then went on to describe Julia's failures:

Cyril: You had the task to establish what needed to be done to meet the organisation's needs. You did not do it so I had to get an expert in to do it for you.

Julia recognised that the real issue was that she had not come up with the solution that Cyril wanted. As he was not prepared to accept her professional view he had to find someone who gave him the answers he wanted.

Julia: I gave you a proposal to resolve this issue but you would not accept it and did not give me the authority to go ahead with it. You cannot say that I have done nothing.

Julia's mind was racing; Cyril did not want her to contribute to this session, only to agree that she had failed. She was not prepared to accept statements that were so clearly untrue.

Cyril: Do you have anything to say? You are just being sullen.

Julia had come to the conclusion that there was no point in saying anything and so sat in silence.

Cyril: It is now 9.30. Do you want time to recover?

Julia thought 'why is he saying this when he does not really care about me?'

Julia: No, I am fine, thank you.

Julia was feeling drained. This had been a dreadful experience. She knew that she was worth more than this. There was no attempt to see her point of view at any stage. She had to attend another meeting within five minutes. She was feeling distressed and phoned her husband to arrange a session with her GP. Julia saw her GP later that day. The GP confirmed that although she was psychologically and medically sound, she was suffering from extreme shock that he related to the nature of the meeting.

Julia experienced a number of other such meetings with Cyril and eventually was forcibly removed from her post. Two years later Julia is happily self-employed in her own field of expertise. During the first-year period she experienced a lot of distressing flashbacks to the incidents of the previous year. She still feels vulnerable and reminders of the experiences can trigger strong emotions. She holds the organisation responsible for what happened to her as she had reported what was happening to senior management in the organisation but no action was taken to investigate Cyril's behaviour.

The moral of the story

Julia has learned a lot from her experiences including:

- She is stronger than she thought she was.
- Sick organisations protect bullies.
- It takes time to recover.
- The more competent you are, the greater the target you become.
- She is a survivor.

Re-telling the stories

The four stories describe the experiences of people who have gone through a period of bullying and emerged as a survivor. In each case the storyteller had tried to deal with a bullying situation using familiar coping methods, but to no avail. The intensity of the distress experienced by the storytellers was not just a result of the bullying but also from the loss of belief in themselves as capable and competent people. As stories were unfolded hidden hurts were revealed. Some of the hurts were familiar but a few were unexpected with their meaning hidden, even from the storyteller. With encouragement each storyteller was able to combine the elements of their story in their unique way and then to present their experience of bullying in a way that was consistent with their values, beliefs and social context. Each of these storytellers is a survivor and while the pain of the bullying can never be totally forgotten each has emerged with new learning and self awareness which can help them deal with the challenges of the rest of their lives.

The survivor's psalm

This psalm by Frank Ochberg has offered hope to many people who have experienced violence and bullying:

> I have been victimised.
> I was in a fight that was not a fair fight.
> I did not ask for the fight. I lost.
> There is no shame in losing such fights, only in winning.
> I have reached the stage of survivor and am no longer a slave of victim status.
> I look back with sadness rather than hate.
> I look forward with hope rather than despair.
> I may never forget, but I need not constantly remember. I *was* a victim.
> I *am* a survivor.
>
> Ochberg, 1993

References

Leymann, H. and Gustafsson, A. (1996) 'Mobbing at work and the development of post-traumatic stress disorders', *European Journal of Work and Organizational Psychology*, 5, 2, 251–75.

McAdams, D. P. (1985) *Power, Intimacy and the Life Story: Personal Inquiries into Identity*, Homewood, IL: Dorsey Press.

McLeod, J. (1996) 'The emerging narrative approach to counselling and psychotherapy', *British Journal of Guidance and Counselling*, 24, 2, 173–84.

Ochberg, F. (1993) 'Posttraumatic therapy' in J. P. Wilson, B. Raphael (ed.) *International Handbook of Traumatic Stress Syndromes*, New York: Plenum Press.

Payne, M. (2000) *Narrative Therapy: An Introduction for Counsellors*, London: Sage.

Scott, M. J. and Stradling, S. G. (1994) 'Post-traumatic stress disorder without the trauma', *British Journal of Clinical Psychology*, 33, 71–4.

Tehrani, N. and Westlake, R. (1994) 'Debriefing individuals affected by violence', *Counselling Psychology Quarterly*, 7, 3, 251–59.

White, M. (1998) 'Saying hullo again: the incorporation of the lost relationship in the resolution of grief' in C. White and D. Denborough, (eds) *Introducing Narrative Therapy*, Adelaide: Dulwich Centre Publications.

Wilson, J. P. (1989) 'Culture and trauma: the sacred pipe revisited' in J. P. Wilson, *Trauma Transformation and Healing: An Integrative Approach to Theory and Research and Post Traumatic Therapy*, New York: Brunner Mazel.

Wingard, B. (1998) 'Grief: remember, reflect, reveal' in C. White and D. Denborough (eds) *Introducing Narrative Therapy: A Collection of Practice-based Writings*, Adelaide: Dulwich Centre Publications.

PART TWO

Size of the problem

CHAPTER FIVE

Social psychology of bullying in the workplace

CLAIRE LAWRENCE

Introduction

This chapter will examine some of the theoretical bases of bullying behaviour based around an examination of the psychological literature on aggression. Although it does not intend to look at personality or biological determinants of bullying behaviour (see, for example, Randall, 1997), it attempts to match psychological theories with the patterns found in the behaviour of those who bully and those who are bullied. The chapter will, where possible, draw parallels between the study of bullying and the study of aggression in general by highlighting the areas in which the two can be linked and those in which the two phenomena remain distinct. The chapter will focus on social interactionist theories of aggression (see for example Felson and Tedeschi, 1993; Lawrence and Leather, 1999a) in particular and asks the question: can an understanding of bullying behaviour be gained by using this approach?

Some definitional issues

Like so many psychological phenomena, definitions of both bullying and aggression are many and varied. Each definition depends crucially upon the perspective taken by any one author. This section will look at definitions of aggression in order to examine the overlap and distinctions between bullying and aggressive behaviour.

Early views of aggression focused on instincts within the individual which drove him or her to act in a hostile and possibly violent manner (Lorenz, 1966). Other approaches paid more attention to the behavioural component of aggressive acts (Buss, 1961). In such cases, aggression was defined as the delivery of noxious stimuli from one organism to another. The understandings and perceptions of those actors involved in the interaction were afforded little or no explanatory value. Similarly neither the situation in which the aggressive act takes place, nor the participants' understanding of it, are paid attention in behaviourist models. From a social interactionist perspective, however, it is precisely these understandings and perceptions that are central in the explanation of violent and aggressive behaviour. The intentions, expectations, beliefs and judgements of those involved are given priority in such explanatory models, together with the prevailing environmental context which provides a setting in which the behaviour is framed (Leather and Lawrence, 1995; Lawrence and Leather, 1999b).

Siann (1985) offers a useful definition of aggressive behaviour. She maintains that four conditions must apply in order for an act to be deemed aggressive. First, the person intends to carry out the behaviour, therefore an action that results in accidental harm cannot be described as aggressive. Second, the behaviour takes place within an interpersonal situation characterised by conflict or competition. It could be argued that this conflict or competition is a subjective perception of the situation and would involve at least one of the actors understanding the context of the act in such a manner. Third, the behaviour is performed intentionally to gain greater advantage than the person being aggressed against does. In this way some form of instrumentality is clear. Fourth, the person carrying out the behaviour has either provoked the conflict or has moved in on to a higher degree of intensity. As a result, there is a sense of initiating or escalating the conflict.

Bullying has also attracted a variety of definitions. Rayner and Hoel (1997), for example, define bullying within five main categories:

1. *Threats to an individual's professional status* – in the work context, this may include public humiliation, having ideas derided, accusation of mistakes;
2. *Threats to an individual's personal standing* – this may include insults and teasing, or spreading rumours;
3. *Isolation* – this could involve withholding work-related information or prohibiting access to opportunities for development;
4. *Overwork* – this would include more than simple high workload as part of the job. Instead this may involve the setting of impossible-to-meet deadlines or extreme pressure to produce work;
5. *Destabilisation* – this can include a lack of recognition or reward for good work, removal of responsibility, and changing remits of the individual's work.

Rayner and Hoel (1997) also maintain that the victim must actually feel harassed by these activities (see also Lyons, Tivey and Ball, 1995) and that their work must be affected as a result. Additionally, they stipulate that bullying must be a repeated and frequent activity. From this series of categories, the focus is predominantly on the victim and their subjective experience of the bullying behaviour. This causes a problem immediately in the understanding of bullying from a social interactionist perspective as this approach relies on the understandings, perceptions and beliefs of all the parties involved in understanding the behaviour and interaction taking place.

Most research has focused upon the perceptions and accounts of victims rather than those of the bullies themselves. The second half of the interaction therefore can only be surmised using relevant theoretical approaches. Similarly, as most of the work on bullying has focused on school bullying, many definitions have arisen from this scenario. Nevertheless, similar patterns can be applied to the context of workplace bullying. For example, according to Olweus (1980), 'A child [worker] is being bullied or victimised when he or she is exposed, repeatedly and over time, to negative actions on the part of one or more children [colleagues]'. This definition emphasises an aspect of bullying which can be problematic in explanations and understanding of the behaviour. That is, it emphasises the systematic and repeated nature of the behaviour.

While bullying is often a repeated activity, it is also important to consider the impact of single bullying episodes on the victim. Randall (1997) for example, takes issue with including repeated behaviour in any description of bullying. He suggests that bullying behaviour may only occur once but be of such intensity for the victim that the impact of the episode continues to affect their work and their interactions with the bully. The fear of future attacks may be sufficient alone to have a repeated and negative impact upon the victim (Randall, 1997, see also Hoel and Cooper in Chapter 1 of this book). Randall (1997: 4) therefore defines bullying as 'the aggressive behaviour arising from the deliberate intent to cause physical or psychological distress to others'. This definition, however, provides no real distinction between purely aggressive behaviour and behaviour which characterises bullying specifically. This definition would equally describe a situation where two men start to fight over the relative merits of a referee's decision at a football match. Are we to therefore say that all violent and aggressive behaviour is bullying? Throughout the remainder of this chapter it will be suggested that the two are not strictly the same and that there are particular aspects of bullying behaviour that depart from the typical escalatory nature of a violent or aggressive episode. These differences and similarities will be discussed in more detail from the perspective of social interactionist and socio-cognitive approaches to aggression (Novaco and Welsh, 1989; Felson and Tedeschi, 1993; Lawrence and Leather, 1999b).

Aggression and bullying: similarities and differences

According to Lawrence and Leather (1999a; 1999b), the interpretation of an activity as aggressive entails complex social judgements made about the behaviour and the underlying intentions and motives driving the behaviour. The interpretation is contextualised by the prevailing norms and expectations inherent in the social situation. The process of identifying intentions is a highly subjective one but it is crucial if we are to determine whether, for example, we are being aggressed against or not. In the same way, the interpretation of an act as bullying

Figure 5.1 Deciding how to act: cognitive and emotional responses (adapted from Novaco and Welsh, 1989)

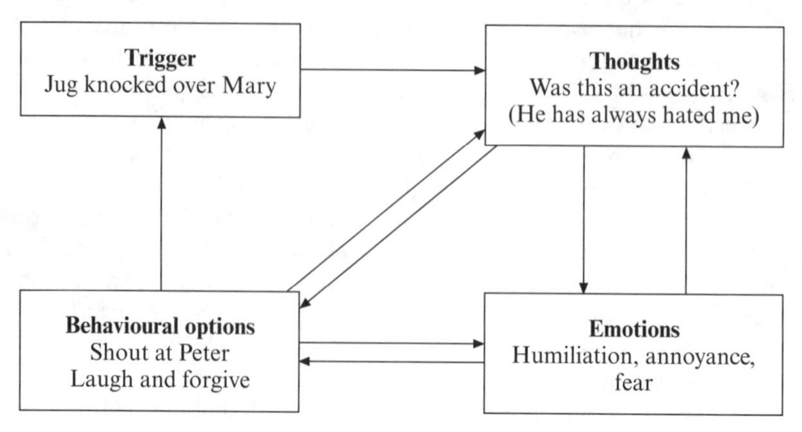

behaviour is also crucial in understanding the dynamics of the interaction and important if we are to determine whether we are being bullied or not. Consider the following example:

Peter goes into the staff canteen with his colleague John. They are sharing a joke when Peter reaches over to pick out a knife and fork. As he does so he knocks a jug of water over Mary who is standing behind him in the queue. Depending upon how Mary interprets Peter's actions, she may decide to react in different ways. Zillman (1978) for example has shown that when people interpret the behaviour of others as malicious, they are more likely to retaliate. However, the work of those studying bullying as a phenomenon points to a lack of retaliation on the part of many victims (Rayner, 1997), even when malicious intent is identified. This is one possible distinction between the two phenomena, and this theme will be examined in more detail later in the chapter. Novaco and Welsh (1989) indicate that we must consider whether a behaviour was intended, justified or serious in its outcome following an incident such as the example above. In doing this, they provide a useful model to illustrate the cognitive stages the victim may pass through (see Figure 5.1).

In this way, the framework offered by Novaco and Welsh (1989) fits nicely into an understanding of what occurs during an incident such as this. In this framework, a trigger (in this case the jug knocked over Mary) results in the victim (Mary) engaging in a series of thoughts about what has happened with the accompanying feelings that these thoughts provoke (for example, anger, embarrassment and exasperation). The target then has to decide what to do about the incident and how, or whether, to respond to the actor (Peter). These are the behavioural options that, according to Lawrence and Leather (1999a), will depend upon the prevailing norms of the situation and the personalities of the participants. As personality determinants of bullying behaviour have been examined elsewhere in this volume, this chapter will examine the situational and interaction-based variables in more detail.

Placing the incident in its context

According to the social interactionist point of view, aggressive incidents can be understood in the context of the entire social episode in which they occur. The setting in which the aggression takes place therefore has an explanatory role in understanding the way in which the behaviour is shaped and perceived. For example, in crude terms, tackling someone to the ground is perceived as being within the normal rules of a game of rugby but not within the context of an academic debate during a seminar. Similarly, some organisational cultures will maintain an adversarial and aggressive approach to work and inter-personal relationships that may support aggressive interactions between colleagues, management and staff, and staff and the clients they deal with (Hoel and Cooper, 2000). While this type of culture may not aim to generate a by-product of hostility and aggression, the effects can be damaging for the organisation as a whole. The important point is that individuals behave according to prevailing rules and expectations. Therefore environments that support a degree of hostility may potentially legitimise aggression and bullying behaviour. This position has been supported by a study conducted by Hoel and Cooper (2000) in which 5,300 workers were surveyed to examine the incidence of bullying and the context in which bullying at work takes place. As well as revealing that bullying behaviour takes a variety of forms including abuse, imposition of impossible deadlines and persistent fault finding, the study suggested that bullying is used by managers to deal with competitive and increasingly demanding work situations. Such competitive environments can arise from job insecurity and, indeed, the authors suggest that a culture in which job insecurity prevails can foster bullying.

The importance of organisational culture and the prevailing social rules are highlighted in contemporary models of aggression. The prevailing culture of an organisation can, for example, be understood as a series of norms and understandings that are shared in some way. Just as organisational cultures are often referred to as 'the way we do things around here' (Deal and Kennedy, 1982), norms are commonly understood as expectancies about what an individual should or should not do in a particular situation (DeRidder et al., 1992).

Norms for behaviour

Norms can be inferred from the behaviour of individuals within an organisation, particularly if those individuals are strongly associated with the organisation, for example a manager or 'founder'. In the same way, there are norms that govern the social interactions that develop as a function of the work situation. As much behaviour, language and materials have symbolic meaning for humans, it is not surprising that norms are frequently communicated symbolically. For example, the design and décor of an organisation's reception area has been shown to communicate the culture and norm systems prevalent within the workplace (Ornstein, 1992). However inferred, knowledge of the norm or rule system of any organisation is vital in evaluating two aspects relating to aggressive and bullying behaviour. First, if the normative culture is characterised by competition and hostility (Hoel and Cooper, 2000), aggressive and bullying behaviours are more likely to be perceived as a legitimate means to obtain goals (cf. Siann, 1985). Indeed as Hoel and Cooper (see Chapter 1 of this book) discuss, when levels of bullying are exhibited from the top of the organisation down through the ranks, it is likely that bullying tactics are also being 'cascaded' downwards, thus perpetuating the bullying norm. Second, knowledge of any norm system is vital in evaluating whether a rule has been broken and identifying the 'injured parties'. In many escalatory aggressive encounters, it is this component that can lead to a perception of wrongdoing resulting in retaliatory behaviour on the part of the victim or the victim's representative (Tedeschi and Nesler, 1993; Lawrence, 1998). In escalatory models of aggression and violence (excluding bullying behaviour), then retaliation or redressing behaviour can be seen in the context of a developing aggressive episode as shown in Figure 5.2 below (following Novaco and Welsh, 1989, developed by Lawrence and Leather, 1999a; Cox and Leather, 1994).

This approach utilises Novaco and Welsh's model (1989) and contextualises interpersonal dynamics with the social and physical environment. The model points out that following an initial event which may have been sparked by some triggering factor (time pressures, innate hostility and so on), both participants in the interaction will consider what to do next. If person 1 decides that the triggering event is the responsibility of person 2, and that person 2 has been malicious and intentional in making the event occur then, in an escalatory model, person 1 may have feelings relating to anger, frustration, etc. These feelings lead person 1 to make decisions about what to do about the situation. If he or she decides to retaliate in some way, then the actions of person 1 become the trigger for person 2. Person 2 then goes through the same process and the incident escalates, with both parties attempting to 'win' as the other 'loses'. In such models, there is the assumption that both parties are relatively well matched in terms of the variables important for that encounter. If the interaction is likely to become physically violent, then a 'match' relating to physical power will be important. If the interaction is related to social positioning, then a 'match' in terms of status is assumed. However, in bullying situations, the presence of 'matched' adversaries cannot be assumed. This is particularly the case when it is considered that 75 per cent of bullies are likely to be their victim's manager (Hoel and Cooper, 2000).

The escalatory model also assumes retaliation as a chosen behavioural option. However retaliation is not the only means open to the victim and, in the

Figure 5.2 Escalatory pattern of aggressive incidents (Lawrence and Leather, 1999)

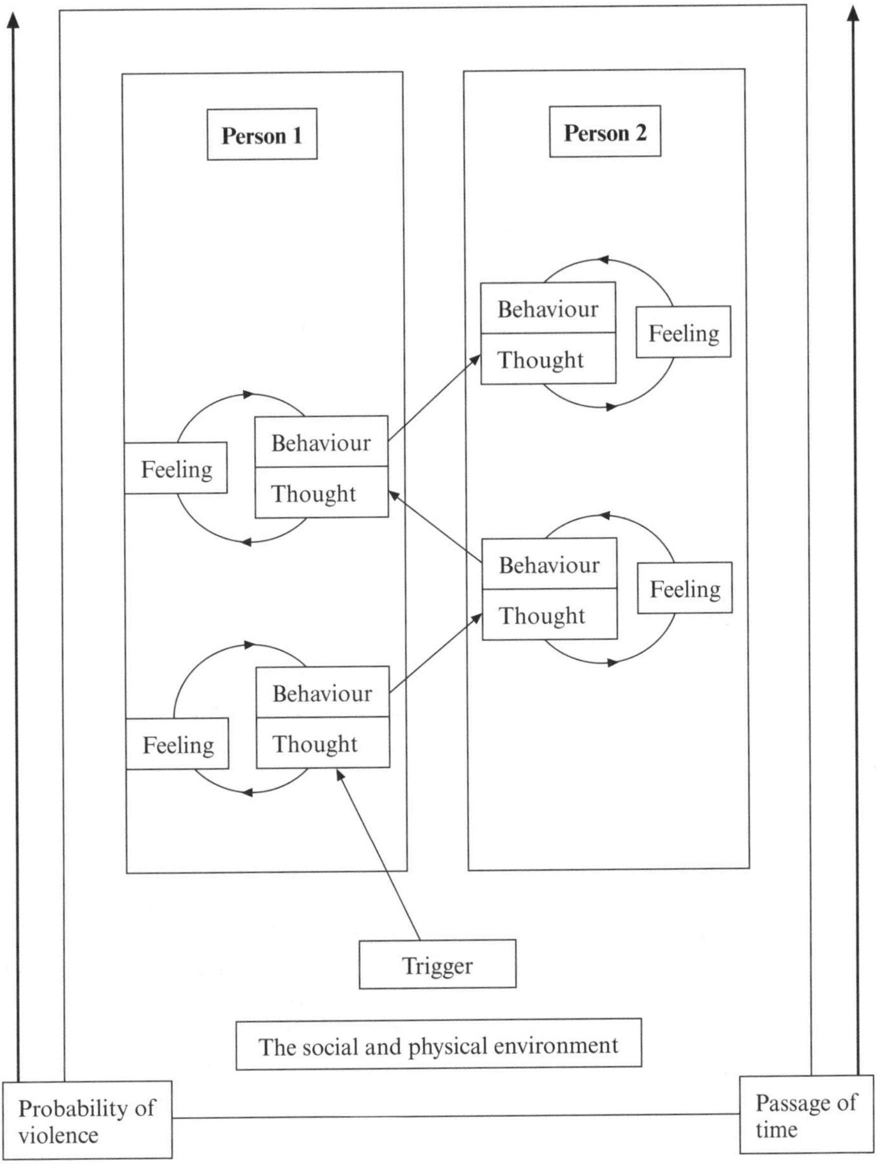

Person 1
Person 2
Behaviour
Thought
Feeling
Behaviour
Thought
Feeling
Behaviour
Thought
Feeling
Behaviour
Thought
Feeling
Trigger
The social and physical environment
Probability of violence
Passage of time

bullying context, retaliation may not take place owing to a variety of forces. Tedeschi and Nesler (1993) outline a range of alternatives open to a victim of aggressive behaviour:

1 The victim may reappraise the norm violating behaviour. For example, the victim may excuse the behaviour as merely the result of the excessive demands

of the situation if the action was seen as the result of a legitimate goal. Using the example earlier, this could occur if Mary notices that Peter spilled the jug over her while reaching to stop a tray of cutlery falling on a child below. In such cases, the offending person may be forgiven, thus permitting the victim to reject other forms of redress that may not be advisable in the situation. However, bullying behaviour is not usually the result of a person acting for the common good. The victims may, at least initially, not speak out or deal directly with the bully, for fear that they will be perceived as not 'playing the game' or not helping the team to achieve its goals. If the case of the bullying boss is considered, the victim may believe that attempting to comply with unreasonable deadlines may be acceptable within the overall goal of the organisation, particularly in periods of high job insecurity. Once again the prevailing norms may have some influence here, particularly if the victim witnesses others tolerating the same behaviour.

2 The victim may decide to demand some redressing action from the wrongdoer. A refusal to comply with this demand may intensify the conflict, further resulting in an escalation into a more aggressive encounter.

   However, many victims of bullying fear this escalation and are concerned about the escalatory nature of the interaction. This is particularly the case if there is a power imbalance between the bully and the victim. This is a common scenario (Rayner, 1997). In the example earlier, if Peter was Mary's boss, then Mary may be less willing to demand an apology as Peter would be in a good position to escalate the hostility further, perhaps by denying Mary's promotion application or humiliating her at the next group meeting. Once again, knowledge of the personalities involved and the culture of the organisation will assist Mary (the victim) in deciding whether or not it is 'safe' to pursue this course of action. Randall (1997: 5) illustrates this fear of retaliation by using the following case study: 'Some really large nurse manager got the better of me he kicked me in the stomach. Next he put a needle against my eye and told me that if I didn't obey his every order then I knew what to expect. That was the one and only time he bullied me but I made a point of never crossing him ' (Randall, 1997: 5). In this example, it is clear that the potential power of the bully has dissuaded the victim from taking redressing action. According to Tedeschi and Nesler (1993), this is a real possibility following attack. Thus, if the victim believes that any redressing action will have no effect or actually make matters worse, then it is more likely that no action will be taken.

3 The victim may decide to punish the 'wrongdoer'. The aim of the punishment is usually to deter the person from performing similar actions in the future and/or to redress the power relationship between the two parties.

In a study by Rayner (1997), the differences in victims and non-victims of bullying are seen to differ in this respect. The difference is illustrated thus: 75 per cent of those who had not been bullied said that they would use this strategy, while only 45 per cent of those who had been bullied said that they would. This difference could be accounted for by the fact that those who had been bullied had experience of a bullying encounter and so are giving a more realistic picture of what they would do. It may be that those who had not been bullied may have a change of heart if they ever became victims of bullying. Nonetheless, these differences in figures may illustrate a difference between victim and non-victim and perhaps help understand why bullies target the individuals they do. In this way it is possible to take into account how aspects of the victim's position in the organisation, their own personality characteristics, their position relative to the bully or the presence of a supporting bullying culture can influence the experience of the victim and their likely response. In particular the issue of power and instrumental aggression is an important consideration for those studying the process of bullying.

Power, instrumentality and bullying

Power – or rather the existence of a power imbalance – often appears in definitions of bullying (Olweus, 1991; Bjorkqvist, 1994). Brodsky (1976) also perceives bullies as manipulating their colleagues or staff in order to achieve power or privilege, pointing out that some positions of power include the remit to inflict actions on others which could be perceived as aggressive (the example given is a Marine Corps drill instructor). For this reason it can be seen that, for some managers, the use of bullying tactics to achieve organisational goals is simply perceived as a means to an end. As discussed elsewhere in this volume, this kind of approach has serious flaws, as a bullying style of management can have a range of negative impacts on the organisation and on those within it. The use of bullying to achieve some goal indicates the extent to which bullying behaviour can be seen as an instrumental activity.

It is worth examining this idea further with reference to the literature on aggression. In many published books and articles about aggressive behaviour and violence, the dichotomy between instrumental aggression and affective (sometimes termed emotional) aggression is made (see for example, Berkowitz, 1993). The difference between the two types of aggression is traditionally understood in the following manner. Instrumentally aggressive behaviour may be viewed as that which is deliberately chosen, planned and employed with the primary purpose of achieving a specific goal. Here, the aggressive act is used merely as a means to an end. A common example of instrumental aggression is the bank robber who uses force or the threat of force in order to achieve their primary goal of financial gain. Affective aggression, however, is usually described as behaviour which is not planned necessarily, but which is rather a response to an unpleasant or stressful situation or encounter (see Berkowitz, 1993). Using our example earlier, if Mary lashed out at Peter for spilling the jug over her, that could be considered an act of affective or emotional aggression because Mary was reacting in response to the event and was not aiming to gain anything by her actions. However, this last point is a contentious one. Does Mary really gain nothing from responding aggressively in response to Peter's behaviour? According to the social interactionist framework, the beliefs, expectations and normative environment of the individuals involved are of paramount importance. Knowledge of these beliefs and perceptions can help us to understand why aggression was chosen above all other options, for example, laughing the situation off, allowing an apology to be made, etc. The blame attributed to individuals for aggressive incidents may crucially depend on an understanding of the motivational basis of the 'aggressor's' behaviour. In this way it is important to consider that aggressive actions which may at first glance appear emotionally driven often, in fact, contain some instrumental elements (Lawrence and Leather, 1999a).

Returning to the earlier example, Mary may indeed have some instrumental reasons for acting aggressively towards Peter in the circumstances. If Peter and Mary have a history that involves Peter playing unwanted practical jokes on Mary, she may take this opportunity to indicate her annoyance and anger and therefore dissuade him from acting in a similar way in the future. She may, alternatively, be worried that she is too placid and that people take advantage of her. She may react aggressively, therefore, as a means of displaying that she is a force to be reckoned with. These explanations will involve an understanding of the dynamics of the interaction and so to categorise aggression into purely instrumental or affective motivations is often to limit the complexity of the exchange.

When we turn to examine bullying, the research into the motivations behind bullies is less well developed. However, there are some important points to note. Much of the work on adult and workplace bullying has revealed that bullying behaviour does appear to have a strong instrumental component. Correlations have been found, for example, between bullying and insufficient work control and high levels of role conflict (Einarssen et al., 1994). This implies that bullies may see their activities as maintaining control over their colleagues or staff. As mentioned earlier, Brodsky (1976) has indicated that bullying behaviour may be related to the need to attain power or privilege – either formally by the gaining of reward and promotion, or informally by the power obtained from generating terror among co-workers. According to Randall (1997: 7) 'it is hard to find instances where bullying has not involved an imbalance of power in favour of the bully'; a point supported by Olweus (1993). Therefore as instrumental aggression is usually used to establish or maintain some form of power over others (Tedeschi, 1983), it appears likely that bullying behaviour in particular can be understood instrumentally. In this way, by understanding what it is that bullies gain from their behaviour, an understanding of how to limit their need to bully others can be achieved. Because of this instrumental aspect of bullying, it may be that intention to harm the victim is not the primary goal of the bully in all cases (see Hoel and Cooper in Chapter 1 of this book). Importantly, and related to instrumental aggression, the social learning process of individuals may give some suggestions for the reduction of bullying behaviour. The nature of social learning and its role in aggression and bullying will now be explored further.

Social learning approaches

According to Randall (1997), there is a very good reason why bullies act in the way that they do – they have learned that it pays off. Individuals may begin their bullying 'career' in childhood where they learn that if they hit another child, they can make the child surrender their lunch money/sweets/new toy. In adults this bullying behaviour may become more 'sophisticated' and in the workplace less physically violent methods may be employed. Nevertheless, the principle is similar. The bully has learned that acting in certain ways results in reward. 'Reward' here is a slightly more complex concept and relates to conditioning models – otherwise known as reinforcement. Here a reward is something that increases the likelihood of behaviours being repeated. Therefore a reward could just as easily be the avoidance of a negative outcome. For example, a manager might realise that by emotionally blackmailing employees to work unreasonable hours, he avoids gaining the negative image of being an inadequate manager with a poorly performing team. According to Bandura (1978), this process is vitally important, particularly in the examination of aggressive and antisocial behaviour. Thus while the social interactionist perspective emphasises the importance of 'rules' of social interaction in governing the way in which we deal with grievances and conflict, the social learning approach focuses more on the role of social development. Bandura's model (1978) goes further than simply using classical and operant conditioning models of reward and punishment for behaviour, however. He also maintained that the revelation that aggressive (or bullying) behaviour pays off can occur through direct experience (as in conditioning models) or by observing someone else using aggressive behaviour to achieve some goal. From this observation, Bandura argues that the child can assimilate the complex skills in the art of aggression and bullying by imitating the observed behaviour (or a form of it). This is known as 'modelling'. This approach is quite useful for an understanding of bullying behaviour, particularly in light of the instrumental component of much bullying. It is therefore essential that organisations are conscious of whether or not they are rewarding bullying behaviour through various mechanisms. For example bullying can be rewarded indirectly through the organisational culture (e.g. admiring talk of 'playing hardball' or circulating stories of organisational 'heroes' who were essentially bullies) or more directly through promotional strategies (i.e. promoting individuals because of the results they achieve – without considering how they were achieved).

At first glance, social learning approaches and social interactionist models may appear to contradict each other somewhat. That is, social learning could be viewed as a strictly behaviourist approach to bullying and aggression without the need to consider the cognitive processes occurring during the interaction. It is these cognitive processes which are so crucial to the social interactionist perspective. However, Bandura later adapted his social learning theory in the light of an increased emphasis on the individuals' interpretative processes. As a result, he indicated (Bandura, 1983) that a 'self-regulatory' mechanism is established within individuals which is the result of a set of learned, directly experienced standards of behaviour. This mechanism will mean that the individual can experience displeasure (or shame) at their own behaviour – even in the light of no external punishment. This displeasure is sufficient to prohibit the aggressive or antisocial action.

Group bullying and witnessing bullying

Finally, group bullying and the role of witnesses of bullying behaviour will be considered. In particular, two processes will be discussed: de-individuation and bystander apathy.

De-individuation: accountability and the influence of group

As discussed earlier, norms are very important in providing standards of behaviour in any given environment. In most societies, antisocial and aggressive behaviour is anti-normative and therefore one of the reasons that individuals do not engage in antisocial behaviour is the fear of social disapproval or condemnation. If this fear is removed, then it follows that the drive to uphold societal norms is reduced. Zimbardo (1970) described this process and de-individuation. It primarily involves a reduction in the extent to which an individual monitors their behaviour and a reduction in their concern about the way in which others perceive them. De-individuation can occur:

- by an individual being anonymous to those present, which can reduce the possibility of being blamed and punished;
- by an individual being a member of a group, which can increase anonymity and therefore reduce the likelihood of them shouldering the full force of blame and responsibility;
- by the situation affording few rules for appropriate behaviour, which means that the norms which would govern behaviour are less influential;
- by an individual being under the influence of drugs or alcohol, which allows a person to later disassociate themselves from their behaviour while intoxicated.

Deiner (1980) has also suggested that when an individual acts as a member of a group, they become less self-reflective and instead look to the group to provide the inhibitory cues for aggressive or antisocial behaviour. If no such cues are available, or if the cues actually promote such behaviour, then the individual may be more disposed to follow a group-generated norm. As group bullying is relatively common (in a study by Rayner (1997), 38 per cent of the sample had been bullied by a group of more than five people), it may be that some de-individuation is occurring. It may also serve to reinforce an individual's positive self-image within the bullying group.

Bystander apathy

Examining bystander apathy, it is surprising to note that in Rayner's study (1997), 77 per cent of the sample had witnessed bullying at work. It is likely given this number that many individuals witness bullying and yet do nothing to confront the bully or step in to help the victim. Although this may appear a somewhat callous neglect of the well-being of their colleagues, there may be many reasons for the inaction. Firstly, if the bully is a manager, which they often are (Hoel and Cooper, 2000), then the witness may fear for their own position within the organisation following retributional action by the bully. Secondly, the witness may simply fear becoming a victim of bullying. Thirdly, Rayner (1997) also reports that a recent job change on the part of the victim accounts for 51 per cent of the incidence of bullying. Therefore it is likely that, because the victim is new to the work group, they have had insufficient time to form support networks. As a result, a shared responsibility for the victim amongst the colleagues can result in no one colleague taking action on behalf of the victim. Organisations need to be particularly vigilant for bullying behaviour in such situations. Rayner (1997) also highlights the change of manager as a key period for bullying to begin. These periods of change therefore need to be identified as potential bullying 'hot spots' and organisational awareness should be increased around these times in order to reduce the incidence of bullying and the negative impact of bystander apathy.

Conclusions and organisational issues

This chapter has outlined the social interactionist approach to violence and aggression and indicated the extent to which this is an appropriate model to understand bullying. The chapter has suggested that the need to understand the cognitive bases for conflict and bullying should not be overlooked, but that the traditional escalatory nature of some models (Cox and Leather, 1994, for example) does not address the power imbalance which typifies many bullying episodes. The chapter has also emphasised the instrumental nature of much bullying behaviour and indicated that it is necessary to understand the gains of bullying for the bully. In this way, organisations need to be careful not to reward bullying behaviour as this will not only serve to reinforce the bully's actions and reduce staff morale but also send a clear message to others in the organisation that bullying is acceptable. The chapter has also discussed the incidence of bystander apathy and therefore organisations should offer ways for staff to report bullies without fear of retribution or damaging their career prospects. The organisation has a leading role to take in the reduction of bullying and it is essential that measures to combat bullying be implemented across every level. It is essential that a total and integrated organisational approach to bullying is accomplished in order to reduce its incidence and thereby the misery experienced by the victims of workplace bullies.

References

Bandura, A. (1978) 'Learning and behavioural theories of aggression' in I. L. Kutash, S. B. Kutash, L. B. Schlesinger and associates (eds) *Violence: Perspectives on Murder and Aggression*, San Francisco: Josey-Bass.

Bandura, A. (1983) 'Psychological mechanisms in aggression' in R. G. Geen and E. I. Donnerstein (eds) *Aggression: Theoretical and Empirical Issues*, New York: Academic Press.

Berkowitz, L. (1993) *Aggression: Its Causes, Consequences and Control*, New York: McGraw-Hill.

Bjorkqvist, K. (1994) 'Sex differences in aggression', *Sex Roles*, 30, 177–88.

Brodsky, C. M. (1976) *The Harassed Worker*, Toronto: Lexington Books.

Buss, A. H. (1961) *The Psychology of Aggression*, New York: Wiley.

Cox, T. and Leather, P. (1994) 'The prevention of violence at work' in C. L. Cooper and I. T. Robertson (eds) *International Review of Industrial and Organisational Psychology*, Chichester: Wiley and Sons, 9, 213–45.

Deal, T. W. and Kennedy, A. (1982) *Corporate Cultures: The Rites and Rituals of Corporate Life*, Reading, MA: Addison-Wesley.

Deiner, E. (1980) 'Deindividuation: the absence of self-awareness and self-regulation in group members' in P. Paulus (ed.) *The Psychology of Group Influence*, Hillsdale, NJ: Erlbaum.

DeRidder, R., Schruijer, S. G. and Tripathi, R. C. (1992) 'Norm violations as a precipitating factor of negative inter-group relations' in R. DeRidder and R. C. Tripathi (eds) *Norm Violation and Inter-group Relations*, New York: Oxford University Press.

Einarsen, S., Raknes, B. I. and Matthiesen, S. B. (1994) 'Bullying and its relationship to work and environment quality: an explanatory study', *European Work and Organisational Psychologist*, 4, 381–401.

Felson, R. B. and Tedeschi, J. T. (eds) (1993) *Aggression and Violence: Social Interactionist Perspectives*, Washington DC: American Psychological Association.

Geen, R. (1990) *Human Aggression*, Milton Keynes: Open University Press.

Hoel, H. and Cooper, C. L. (2000) 'Workplace bullying in Britain: some key findings from a study of 5,300 employees', *Employee Health Bulletin*, 14, 6–9.

Lawrence, C. (1998) 'Forming impressions of public house violence: stereotypes, attributions and perceptions', unpublished doctoral thesis, University of Nottingham.

Lawrence, C. and Leather, P. (1999a) 'The social psychology of aggression and violence' in P. Leather, C. Brady, C. Lawrence, D. Beale and T. Cox (eds) *Work-related Violence: Assessment and Intervention*, London: Routledge.

Lawrence, C. and Leather, P. (1999b) 'Stereotypical processing: the role of environmental context', *Journal of Environmental Psychology*, 19, 383–95.

Leather, P. and Lawrence, C. (1995) 'Perceiving pub violence: the symbolic influence of social and environmental factors', *British Journal of Social Psychology*, 34, 395–407.

Lorenz, K. (1966) *On Aggression*, New York: Harcourt Brace Jovanovich.

Lyons, R., Tivey, H. and Ball, C. (1995) *Bullying at Work: How to Tackle It*, a guide for MSF representatives and members, London: MSF.

Novaco, R. W. and Welsh, W. N. (1989) 'Anger disturbances: cognitive mediation and clinical prescriptions' in K. Howells and C. R. Hollin (eds) *Clinical Approaches to Violence*, Chichester: John Wiley and Sons Ltd.

Olweus, D. (1980) 'Familial and temperamental determinants of aggressive behaviour in adolescent boys: a causal analysis', *Developmental Psychology*, 16, 644–60.

Olweus, D. (1991) 'Bully/victim problems among school children' in I. Rubin and D. Pepler (eds) *The Development and Treatment of Childhood Aggression*, Hillsdale NJ: Erlbaum, 411–47.

Olweus, D. (1993) *Bullying at School: What We Know and What We Can Do*, Oxford: Blackwell.

Ornstein, S. (1992) 'First impressions of symbolic meaning connoted by reception area design', *Environment and Behaviour*, 24, 85–110.

Randall, P. (1997) *Adult Bullying: Perpetrators and Victims*, London: Routledge.

Rayner, C. (1997) 'The incidence of workplace bullying', *Journal of Community and Applied Social Psychology*, 7, 199–208.

Rayner, C. and Hoel, H. (1997) 'A summary review of literature relating to workplace bullying', *Journal of Community and Applied Social Psychology*, 7, 181–91.

Siann, G. (1985) *Accounting for Human Aggression: Perspectives on Aggression and Violence*, Boston: Allen and Unwin.

Tedeschi, J. T. (1983) 'Social influence theory and aggression' in R. G. Geen and E. I. Donnerstein (eds) *Aggression: Theoretical and Empirical Reviews*, New York: Academic Press.

Tedeschi, J. T. and Nesler, M. S. (1993) 'Grievances, development and reactions' in R. B. Felson and J. T. Tedeschi (eds) *Aggression and Violence: Social Interactionist Perspectives*, Washington DC: American Psychological Association.

Zillman, D. (1978) 'Excitation transfer in communication-mediated aggressive behaviour', *Journal of Experimental and Social Psychology*, 7, 419–34.

Zimbardo, P. G. (1970) 'The human choice: individuation, reason and order versus deindividuation, impulse and chaos' in W. J. Arnold and D. Levine (eds) Nebraska symposium on motivation, Lincoln: NE University of Nebraska.

CHAPTER SIX

Monitoring bullying in the workplace

DIANE BEALE

Introduction

This chapter views bullying from an organisational point of view, discussing first what should be considered as work-related bullying, and the extent of the problem as revealed in the academic and professional literature. It then outlines briefly the health and safety approach to tackling bullying, and suggests ways in which the problem might be monitored effectively.

Bullying may produce a number of adverse consequences at an organisational level, such as absenteeism, high staff turnover and under-performance, as well as personal consequences to individual members of staff (e.g. Alderman, 1997; Farrell, 1999; Quine, 1999; Cooper and Hoel, 2000). As such, it is a phenomenon that warrants serious consideration by all employers on economic, as well as humanitarian, grounds. In order for an organisation to monitor and tackle bullying, it is necessary to understand what to look for and to recognise the signs and effects.

What qualifies as bullying?

Bullying and related terms

Bullying in work situations has received increasing attention in the academic, professional and managerial literature over the last decade, but a number of related terms have been used. Einarsen (1999: 17) asserts that the concepts of 'mobbing', 'emotional abuse', 'harassment', 'bullying', 'mistreatment' and 'victimisation' have all been used to describe the same phenomenon, that is 'the systematic persecution of a colleague, a subordinate or a superior, which, if continued, may cause severe social, psychological and psychosomatic problems for the victim'.

While all these terms have been used for very similar phenomena, each has slightly different boundaries and nuances. The term 'victimisation', for example, may be thought of as the most general term. Aquino et al. (1999: 260) describe victimisation as 'an individual's perception of having been exposed, either momentarily or repeatedly, to the aggressive acts of one or more other persons'. The term is also used widely in national statistics relating to workplace violence (e.g. Warchol, 1998) where it may be applied to any violent or aggressive incident including single events that would not necessarily qualify as bullying.

Scandinavian and German researchers have used the term 'mobbing' for the same phenomenon as bullying (e.g. Van Dick et al., 1999; Wolfgang, 1999; Zapf, 1999), and it is defined by Zapf (1999) as 'a severe form of social stressors at work. Unlike "normal" social stressors, mobbing is a long lasting escalated conflict with frequent harassing actions systematically aimed at a target person'. However, it has been argued that mobbing usually refers to groups of people repeatedly attacking, or putting undue pressure on, an individual (Ramage, 1996), certainly the term in English conveys that impression.

The term 'harassment' has tended to be limited, during the past, to behaviours that include a sexual, racial or disability component, in accordance with the sex, racial and disability discrimination legislation (see Chapter 7), although Einarsen (1999) usefully calls bullying 'generic harassment'. The Protection from Harassment Act (1997) has also served to generalise the term harassment by including the concept of the view of the 'reasonable person', namely 'the person whose course of conduct is in question ought to know that it amounts to harassment of another if a reasonable person in possession of the same information would think the course of conduct amounted to harassment of the other'. In this act, harassment is taken to include 'alarming the person or causing the person distress' and requires the behaviour to occur 'on at least two occasions'. Harassment, however, does not necessarily include the use of power by the perpetrator, in contrast to bullying. Harassment and bullying are often used almost interchangeably in the literature and within many organisational policies (Ishmael and Alemoru, 1999).

Another term closely related to bullying is 'petty tyranny', as used by Ashforth (1994: 755). This occurs where an individual 'lords his or her power over others', displaying behaviours that include arbitrariness and self-aggrandisement, belittling subordinates, lack of consideration, a forcing style of conflict management, discouraging initiative and inappropriate punishment. This term refers to a subset of bullying behaviours, conveying the more obvious manifestations of bullying, but not those that are subtler.

The other concept that overlaps with bullying is that of work-related violence, particularly co-worker violence, which has merited much attention in the American literature (e.g. VandenBos and Bulatao, 1996). However, bullying is not always as easily recognised as violence. While it may consist of repeated overt aggression or physical violence, it may equally involve covert, subtly malicious acts that, over time, undermine the victim. This can be thought of as insidious psychological violence through such behaviours as setting unrealistic work targets, giving meaningless tasks, spreading malicious rumours or taking the credit for others' work.

At a more general level, Andersson and Pearson (1999), in describing deteriorating interpersonal relationships at work, use the term 'workplace incivility', defined as 'low-intensity deviant behaviour with ambiguous intent to harm the target, in violation of workplace norms for mutual respect'. Such behaviours are characteristically rude and discourteous, displaying a lack of regard for others. The spiral of incivility, described by Andersson and Pearson, in which two or more people become increasingly uncivil towards each other, is a useful concept for understanding how some types of bullying may originate and develop. However, it is important to understand that it is also possible to bully someone while appearing extremely civil and courteous.

Definitions of bullying

So what is defined as workplace bullying? As Ishmael and Alemoru (1999) note, there is no set definition, however organisations need a working definition for inclusion in policies and communication with staff. Three useful definitions are given here to illustrate those given in the published literature:

Einarsen et al. (1996) state that 'the term bullying refers to a situation in which a person is subjected to negative conduct from co-workers or supervisors over a period of time, such as harassment, offensive remarks, constant criticism or social exclusion. To qualify as bullying the victim must have difficulties defending himself/herself in the actual situation'.

Lancashire County Council (*Employee Health Bulletin*, 1999) uses the definition: 'Bullying is intimidation on a regular and persistent basis which serves to undermine the competence, effectiveness, confidence or integrity of the target. It involves a misuse of power, position or knowledge to criticise, humiliate or destroy a subordinate or colleague.'

The Industrial Society (Ishmael, 1999) suggests a working definition as 'persistent, offensive, abusive, intimidating, malicious or insulting behaviour, which amounts to an abuse of power and makes the recipient feel upset, threatened, humiliated or vulnerable. Bullying undermines a target's self-confidence and may cause them to suffer stress'.

These definitions have a number of factors in common, but also a number of differences. The features that require consideration are:

- What behaviours constitute bullying?
- What is the frequency with which such behaviour has to occur and for how long does it have to persist to qualify as bullying?
- Who might be the perpetrators and the victims and what is the organisational relationship between them?
- What part is played by the intentions of the perpetrators and the perceptions of the victims and of any witnesses to the behaviour?

Bullying behaviours

None of the definitions given above specifically includes physical violence. However, the individual acts that constitute bullying can range from direct and serious physical attack, through threat and verbal abuse, obstructive actions that damage the target's work performance, to taking credit for work actually performed by the target. The actions may be aimed directly at the target or may be effected indirectly via remarks made to others. It can be argued that showing unmerited favouritism to one employee at the expense of others is also a form of bullying. A number of researchers have attempted to classify types of bullying behaviour. Van Dick et al. (1999), for example, have used a measure of bullying which contains four items that concern ignoring people (acting as though they were not there), unfairly criticising work, spreading rumours and leaving people out of meetings and social events.

One of the most useful classifications of bullying behaviours is that derived by Quine (1999) based on the five categories suggested by Rayner and Hoel (1997). These categories are threat to professional status, threat to personal standing, isolation, overwork and destabilisation. Quine's survey (1999) incorporated an inventory of 20 bullying behaviours that fall into the five categories, as shown in Table 6.1. This inventory is not exhaustive but it serves to illustrate here the wide range of behaviours that may constitute bullying.

Table 6.1 Categories of bullying behaviour (taken from Quine, 1999: 230)

Category	Behaviour
Threat to professional status	Persistent attempts to belittle and undermine your work. Persistent and unjustified criticism and monitoring of your work. Persistent attempts to humiliate you in front of colleagues. Intimidatory use of discipline or competence procedures.
Threat to personal standing	Undermining your personal integrity. Destructive innuendo and sarcasm. Verbal and non-verbal threats. Making inappropriate jokes about you. Persistent teasing. Physical violence. Violence to property.
Isolation	Withholding necessary information from you. Freezing out, ignoring, or excluding. Unreasonable refusal of applications for leave, training, or promotion.
Overwork	Undue pressure to produce work. Setting of impossible deadlines.
Destabilisation	Shifting of goal posts without telling you. Constant undervaluing of your efforts. Persistent attempts to demoralise you. Removal of areas of responsibility without consultation.

Baron et al. (1999) have classified workplace aggression as 'overt' or 'covert'. In overt aggression, aggressors make no attempt to conceal their identity or their actions from the target person. Conversely, in covert aggression, aggressors seek to hide their identity or their actions. This classification of behaviours serves as a reminder that, while bullying behaviours may be obviously and directly aggressive, they can also be extremely subtle, so much so that even the target may not realise it is happening.

Frequency and duration of bullying behaviour

One of the criteria often used in definitions of workplace bullying is that the behaviour has to be repetitive or systematic, as can be seen in the three definitions quoted above. Einarsen (1999), for example, emphasises 'repeated and enduring aggressive behaviours'. This requirement is included mainly to differentiate bullying from isolated outbursts or disagreements, and from individual incidents of violence. Undoubtedly, however, a single incident would be considered to constitute bullying if there was an obvious imbalance of power and the effect was sufficiently severe. Threat of job loss if a worker did not carry out a particular task against their will, or a physical beating by a much stronger individual or group of individuals, are possible examples.

Some researchers have specified particular frequency and duration for behaviour to qualify as bullying. Leymann (1992), for example, required the behaviour to be repeated at least once a week for at least six months. While such requirements may be appropriate for large-scale sociological surveys, they are inappropriate for an organisation wishing to tackle bullying as a problem. The earlier such behaviour is detected and eradicated, the better for both the target and the organisation. Serious damage may be done to an individual within a few days or weeks if the bullying behaviour is extreme or intense. Six months is a very long time for somebody to suffer being bullied before the problem is taken seriously. It is also long enough for such behaviour to become accepted as the norm and to spread into the organisational, or at least the team, culture.

An organisational culture that condones, or ignores, a bullying management style, or bullying within teams, not only increases the likelihood that bullying will spread, but also increases the potential for more serious physical violence to occur. Apparently minor problems, either unrecognised by the organisation or unresolved, can build up to a point of revenge or retaliation, as acknowledged by a number of authors (Fox and Levin, 1994; Bies et al. 1997; Andersson and Pearson, 1999; Baron et al. 1999; Bradfield and Aquino, 1999; Cohen et al. 1999). This reminds us that, if bullying is ignored, the target may not be the only person who suffers in the long term, if 'the worm turns' to take revenge on the perpetrator or the organisation. Greenberg and Barling (1999) found that perceptions of injustice and excessive monitoring were significant precursors of aggression against supervisors. Folger and Baron (1996) have similarly related perceived injustice to the occurrence of co-worker violence.

Perpetrators and victims

One dictionary definition of a bully is 'a cruel oppressor of the weak' (*Chambers Concise Dictionary*). This implies that the perpetrator of bullying must hold some sort of power over the victim that prevents them from defending themselves effectively. In a work situation, such power is often that of superior status within the organisation but may also be possession of such things as information, knowledge, skill, access to resources or social position. Thus, bullying may occur from superior to subordinate (downward bullying), between co-workers (horizontal bullying) or from subordinate to superior (upward bullying), at any level of the organisational hierarchy. The existence of upward bullying is borne out in Quine's survey of 1,100 employees of an NHS community trust (1999) in which 12 per cent of those cited as bullies were of less senior status than the target. For example, a subordinate might undermine a supervisor by withholding or distorting information so that the supervisor is made to look incompetent in front of a superior.

However downward bullying appears to be most widespread, as reported in a number of surveys. Both Quine's survey (1999) and Rayner's survey of members of UNISON, the public employee trade union (1997), revealed that, when bullying occurred, it was most likely to be by a manager. Also Bjorkqvist et al. (1994), in a study of harassment among university employees, found that individuals in superior positions harassed more often than individuals in subordinate positions.

However, Baron et al. (1999), in a study of aggression in the workplace, asked the respondents about their own aggressing behaviour, as well as experience of being aggressed against, and found that workers in their sample were most likely to aggress against co-workers and immediate supervisors and less likely to aggress against subordinates. The tendency to aggress against their supervisors and the organisation was related to the degree of perceived injustice, that is the participant's satisfaction with treatment from their supervisor.

In terms of numbers, two-thirds of the bullied respondents in Rayner's UNISON survey (1997) said that just one person had bullied them. The bullying was usually of more than one person, sometimes the whole section or department. Cooper and Hoel (2000) similarly found that the majority of bullied respondents had been bullied along with others, again sometimes the whole group.

In terms of any gender effects, Bjorkqvist et al. (1994) found that female university employees felt that they were significantly more harassed than their male counterparts. In Rayner's survey (1997), the majority of bullying was found to involve cross-gender interactions, i.e. women bullied by men, and men by women. However, Cooper and Hoel (2000) found that, although bullying of women was likely to involve men, the bullying of men was more likely to be by other men. Such results are likely to be dependent on the type of job and the gender mix, as illustrated in Richman et al.'s study (1999) within an American university. Quine's survey (1999) found that most bullying was carried out by women towards women. This finding was not surprising, however, as the workforce was predominantly female. What organisations need to learn from the diversity of these results is that there is no restriction on the situations, or relative positions of the perpetrator and target, in which bullying can occur.

It should also be recognised that bullying can be carried out by clients as well as other employees, for example by pupils against a teacher (Leyden, 1999). Cooper and Hoel (2000) found that around 8 per cent of bullied respondents were bullied by clients. Although this constitutes rather a separate problem, from an organisational point of view, such bullying must also be monitored and tackled.

Intentions and perceptions

Einarsen's requirement for bullying to have occurred (1999) is that either the intention was hostile or that it was perceived as hostile by the recipient of the behaviour. The Andrea Adams Trust definition (*Employee Health Bulletin*, 1999) includes both intentions ('deliberately humiliating' and 'vindictive') and perceptions ('unwanted' and 'offensive'). Similarly, the Industrial Society (Ishmael, 1999) definition of bullying looks at both the intentions of the perpetrator ('persistent, offensive, abusive, intimidating, malicious or insulting behaviour') and the feelings of the target ('makes the recipient feel upset, threatened, humiliated or vulnerable').

Intentionally damaging behaviour usually occurs because the perpetrator expects to gain some personal advantage. Bjorkqvist et al.'s sample of university employees (1994), for example, gave the reasons for harassment as predominantly envy and competition about jobs and status. An organisation has to ensure that its systems and culture do not reward bullying behaviour, rather that such behaviour will be discovered and challenged. Zapf (1999), for example, argues that organisations should not look at the bully in isolation but rather at the whole context in which bullying behaviour occurs, including the organisation, the social system, the perpetrator and the victim. He asserts that one-sided explanations of the causes, such as the personality of the perpetrator, are likely to be inappropriate as many cases arise from a number of interrelated circumstances. The roles of the personality of perpetrators and the perception of victims are discussed in Chapters 1 and 11.

The identification of intentions and perceptions is an aspect of bullying that causes problems for an organisation in recognising and eliminating bullying. It provides good reason to facilitate informal discussion of any problems without a formal complaint having to be made, so enabling misunderstandings to be resolved at an early stage. It is possible for a person to feel bullied without the supposed bully intending, or being aware of, any harm to the target. Different expectations, perceptions, cultural norms or simple oversights can lead to misunderstandings. If these are interpreted as hostile by the apparent target and not discussed with the apparent bully, so that the misunderstanding persists, the feelings of being persecuted can build up. An example of a misunderstanding might be someone's name being inadvertently omitted from a circulation list so that they repeatedly fail to receive information. If their subsequent lack of action or knowledge is then interpreted as laziness or incompetence, it might result in repeated criticism that seems justified to the person making the criticism, but unjustified to the target, appearing to that person as bullying.

The perceptions of the apparent target of bullying are often cited as the most important aspect of bullying, as discussed in Chapter 1. This concentration on perception has occurred for several reasons. First, the significant detrimental effects upon victims are increasingly seen as unacceptable in the work context (Ishmael and Alemoru, 1999). Second, there can be great difficulty in establishing the intent of the apparent bully. Third, for research purposes, it is more practical to obtain information from people who feel that they have been bullied than from those who have carried out the bullying. However, it is important to recognise that the perception of being bullied is far from infallible as a test of whether actual bullying has occurred. The role of misunderstanding and differing expectations has been noted already in people perceiving that they have been bullied where no harm was intended.

Conversely, a person might be targeted by a bully without being aware. Baron et al. (1999) pointed out that covert workplace aggression occurs where aggressors seek to hide their identity or actions. As a result, targets may not realise that some harm they experienced was due to someone trying to harm them or, indeed, they may be unaware of the harm done to them. An example might involve a manager repeatedly and unfairly criticising the work of a subordinate when speaking to higher management, resulting in that subordinate not receiving promotion or other deserved reward. The fact that the subordinate was unaware of the manager's malicious actions does not lessen the bullying nature of the behaviour. From an organisational point of view, this type of behaviour is particularly difficult to uncover and requires vigilance by higher managers and co-workers.

McDonald and Lees-Haley (1996) have pointed out that certain personality disorders, such as paranoia, can cause an employee to misperceive the words or actions of supervisors or co-workers as malevolent or unjustified when in fact they are not. Further, they observe that borderline personality disorder, for example, can also cause employees to manipulate others to 'set the stage for their own victimisation'. McDonald and Lees-Haley suggest that the presence of a personality disorder might also provide an alternative explanation for why a plaintiff or claimant in litigation may suffer severe emotional distress as a result of relatively innocuous workplace events. In the interests of fairness, organisations have to be aware that, occasionally, such situations can occur, but they must not use this possibility as an excuse to dismiss any complaints as fabrication or imagination on the part of the alleged victim. It may also be that there is a personality disorder or mental health problem on the part of the perpetrator that underlies the bullying behaviour. McDonald and Lees-Haley (1996) note this possibility with respect to obsessive–compulsive and narcissistic personality disorders, for example, while over half of Rayner's survey respondents (1997) cited 'mental imbalance of the bully' as a cause of bullying. Every case has to be taken seriously by organisations and investigated with fairness to both parties so that the underlying issues can be ascertained and resolved.

How much bullying occurs?

It is important for organisations to have some idea of how widespread the problem of work-related bullying is, so that they can appreciate the likely extent of bullying within their own organisation. However, different studies have produced very different results, with varying degrees of generalisability depending on how the subject sample was selected and how the information was gathered. The main issue to take into account when considering responses to surveys on a national level, or carried out through trade unions, is that workers who feel they have been bullied are more likely to respond than those who have not, as they have an interest in raising awareness of the problem. This may cause the survey results to be inflated in terms of the percentage of respondents bullied, compared to the population as a whole. Most researchers are aware of this problem and take it into account in reporting their results. Comparison of results from different surveys can be problematic if they use different specifications of behaviour types or different time frames.

The most comprehensive survey, in terms of the population studied, has been carried out by Cooper and Hoel (2000). This survey of a random national sample of workers in Britain found that, on average over the industries surveyed, around 10.5 per cent of respondents reported having been bullied at work during the preceding six months, 1.4 per cent on a daily or weekly basis, 24.4 per cent considered they had been bullied during the preceding five years and 46.5 per cent of respondents had witnessed bullying. Rayner's survey of UNISON members (1997) gave results of a similar order. Two-thirds of respondents had experienced or witnessed bullying at some time; 18.5 per cent of respondents considered that they had been bullied in the previous six months. Being aware of the problems of generalisability and taking other results into account, Rayner estimated that this represented a working average of 14 per cent of UNISON members.

Within particular industries, rather higher results have been obtained. Quine's survey of 1,100 employees of an NHS community trust (1999) found that 38 per cent of respondents reported experiencing one or more types of bullying in the previous year, while 42 per cent had witnessed the bullying of others. This survey achieved a very high response rate of 70 per cent of all trust employees, so responses are likely to be acceptably representative of the whole staff. Also within healthcare, a Community Practitioners' and Health Visitors' Association survey (Durdle, 1997) indicated that 39 per cent of respondents had personally experienced bullying, humiliation and abusive behaviour from managers or from other professionals; 22 per cent stated that they had been bullied directly and privately.

Within higher education, Richman et al.'s survey of employees at an American university (1999) found that substantial numbers had experienced 'generalised workplace abuse' ranging from 52 per cent to 78 per cent of respondents, depending on sex and occupational group. Clerical and service workers reported experiencing higher levels of particularly severe mistreatment than other groups.

The overall picture provided by such surveys as these is that bullying is a widespread phenomenon, more prevalent in some industries than in others. Even taking the lowest percentage of respondents reporting being bullied, i.e. 10.5 per cent (Cooper and Hoel, 2000), there is a significant problem for employers to tackle regarding the welfare of their staff.

What is the best approach for managing bullying?

Work-related bullying clearly impacts on employee health and well-being (e.g. Cooper and Hoel, 2000) so warrants consideration as a hazard manageable under a health and safety approach, in the same way that both work-related stress and work-related violence can be managed (e.g. Cox and Cox, 1993; Leather, et al., 1999). Such an approach requires risk assessment followed by risk reduction, as stipulated by the Management of Health and Safety at Work Regulations (1999).

Risk assessment has to be based on good quality information about the nature and extent of any bullying and whether it is increasing or decreasing. Obtaining such information is not straightforward as all forms of violence and aggression are known to be under-reported (Beale, 1999). In the case of bullying, this is exaggerated because each individual act of persecution might appear too trivial to be reported. Additionally, people may doubt the support they will get, or even fear reprisal if they report.

Risk assessment may need to be carried out for particular categories of staff, such as new, particularly young, inexperienced members of staff, perhaps older people in a young culture, or anyone who is a bit different from the norm. Preventative action is preferable to reaction once a problem occurs.

How can organisations monitor bullying?

Monitoring involves two separate levels of investigation. First, there is the requirement to know the extent and nature of bullying within an organisation in order to provide information for risk assessment. Second, there is the need to discover individual cases and to track how they develop and are dealt with by the organisation.

Neither of these is straightforward. Incident reporting systems, which are the mainstay of monitoring for work-related violent incidents (Beale, 1999), are unlikely to be as successful for bullying. The under-reporting that is a recognised problem for violence will undoubtedly be far more marked for bullying. Making a direct, formal report is a risky strategy for a victim because, as already discussed, by definition bullying involves a perceived imbalance of power, such that the victim feels that the perpetrator has some advantage, usually to do with the organisation.

In order to report, people have to have confidence that their situation will improve, not get worse. However, in Rayner's survey (1997), when people who considered that they had been bullied took some action, significant numbers reported that nothing was done, the complainant was labelled a troublemaker or they were threatened with dismissal. This last was particularly common where a group of workers had made a complaint together. Similarly, Quine's survey of NHS employees (1999) found that two-thirds of the victims had tried to take action when the bullying occurred but most were dissatisfied with the outcome.

To reassure staff, more sophisticated systems integrating support with reporting have to be put in place. Monitoring has to be set within an integrated organisational response to the problem with senior management demonstrating commitment by incorporating bullying into policies for ensuring staff safety and well-being. Systems and procedures to deal with such problems should be set up in consultation with staff and union safety representatives. However, it needs to be borne in mind that bullying should not be seen as a management versus union problem because managers at all levels are as likely to be bullied as other members of staff (Cooper and Hoel, 2000). Information has to be communicated effectively to staff to ensure that they are aware of the management commitment to eradicate bullying and of the available means of taking action. Employees' knowledge that there are fair procedures available to them has been shown to moderate retaliatory tendencies towards the organisation or other employees (Skarlicki and Folger, 1997). All the systems must be designed to provide monitoring information, but without breaching any assurances of confidentiality.

Effective monitoring has to utilise a multi-pronged strategy, gleaning information from as many sources as possible, as is the recommended practice for risk assessment (e.g. Cox and Cox, 1993). Evidence should be obtained relating to:

- the likely risk factors for bullying present in the organisation (or sections of the organisation);
- workers' perceptions of the amount and nature of bullying occurring;
- behavioural and health outcomes that might be measured more objectively, such as absenteeism, turnover, or visits to occupational health practitioners.

As a starting point for estimating the extent of bullying, it is sensible to take note of any national surveys that involve similar workers or industries, such as those carried out on behalf of unions or professional bodies (e.g. Rayner, 1997; Durdle, 1997). Cooper and Hoel (2000), for example, found that workers reported bullying to be particularly prevalent (above 14 per cent bullied in the preceding six months) in prisons, postal and telecommunications industries and the teaching and performing arts. In the NHS, banking, engineering and the fire service, the reported prevalence was 6–14 per cent and, in information technology and the pharmaceutical industry, prevalence was lowest at below 5 per cent. Prevalence can vary between different functions within one industry. Bjorkqvist et al. (1994), for example, found that university employees who were involved in administration and service experienced more harassment than individuals who were involved in research and teaching. Richman et al. (1999) similarly found that university clerical and service workers experienced higher levels of particularly severe mistreatment than other employees. No organisation should assume that it is immune but should consider the possibility that it has around average amounts of bullying as a rough starting figure.

Organisations should then note any risk factors as revealed in research in terms of particularly vulnerable people, situations and organisational changes. Vulnerable people include new, young or inexperienced staff; those working in a group that is predominantly of the opposite sex; and anyone who is different from the norm in any other way, such as by ethnic origin or disability. Particular events that cause extra stress or competitiveness include alteration to the management structure of a group, either through promotion or the introduction of a new manager; increases in workload; downsizing; and organisational restructuring. Further risk factors are aggressive management style and a macho organisational or team culture.

Reporting systems

The reporting and recording of bullying is not a statutory obligation, unless the individual acts amount to violence reportable under RIDDOR '95 – The Reporting of Injuries, Diseases and Dangerous Occurrence Regulations (1995). However, two-strand reporting systems need to be set up that allow staff to make a direct, formal complaint where bullying is blatant and also allow confidential informal reporting to a designated manager or adviser, or to a contracted outside agency, where the problems are less clear cut. Such an informal system allows complainants to obtain advice about appropriate action and support through the action they choose, such as mediation or formal complaint. It also provides an alternative route to cater for the direct line manager either being part of the problem or being ineffective in dealing with it.

The formal system must provide records covering details of all complaints, the related investigations and any disciplinary or remedial action taken. The informal system must provide summary details of complaints sufficient to be useful in terms of risk assessment, but without breaching confidentiality or making complainants identifiable. The advisers must keep confidential records of all complaints so that they can be used as evidence at a later stage, if the complainant agrees.

Examples that were described to Beale and her colleagues (Beale et al., 1998) serve to illustrate some of the barriers to reporting.

Example 1: At an NHS trust, a team leader had been bullying colleagues. No one complained but several left before the management realised what was happening. There seemed to be a conspiracy of silence. People did not really know who to consult and did not think to go to occupational health practitioners. The staff knew the formal procedure but did not have the confidence to go to more senior management to complain because the team leader was a 'very clever manipulative lady'. Staff also felt that the trust had become more of a business over recent years and they had become less sure of what was required by managers. They felt that people cared less than a few years ago, so were not sure what to do. The health and safety adviser acknowledged that bullying was a problem and was very difficult to deal with because some of the perpetrators were 'very powerful people'. However, he asserted that it had to be tackled and the proper channels would be followed if informal methods did not work.

Example 2: A paramedic reported that some qualified paramedics used their rank or their age to intimidate trainees into using techniques that were outdated or not recommended, for example for lifting patients. The trainees tended not to make a fuss because they did not feel confident enough to report incidents of verbal abuse to their manager. The younger they were the more they seemed to be intimidated and it was only when they gained more experience and confidence that they then turned round and answered back. The paramedic felt that colleagues had to be aware of the situation and had to try to bolster the younger person's confidence to deal with the bullying. Not dealing with this type of situation could lead to sloppy procedures being used and possible back injuries occurring. He indicated that incidents of paramedics bullying trainees were normally dealt with at the station without reporting to central management.

Example 3: A union representative stated that often staff did not want to talk to management about something that had happened to them. Although the management told the staff that they could be trusted, some people had been to management in the past, spoken about certain problems and then found that their problems were being talked about in front of other staff.

These examples indicate that any increase in the reporting of bullying may be a sign of increasing confidence in the management attitude to supporting people who feel that they have been bullied, rather than a sign of increased bullying. When management actively promote anti-bullying measures, they should expect an initial increase in reporting and see it as a positive sign. This is reflected in the IRS survey of employers (*Employee Health Bulletin*, 1999) where a group of employers reported an increase in the number of bullying cases brought to their attention following the introduction of a policy to provide employees with a route to raise a complaint of bullying. One local government organisation explained this as 'raising the issue through the introduction of a bullying and harassment at work policy has made employees more confident that a complaint will be dealt with.'

Staff surveys

Staff surveys may be useful in obtaining information about the nature and extent of bullying, as long as the limitations are recognised and the surveys are carefully planned. Simply sending out a short questionnaire labelled 'bullying survey' and asking people whether they have been bullied is unlikely to produce useful or reliable results. People who have had no experience are likely to ignore it or to treat it as a joke, while those who have been bullied may be afraid of being identified and further victimised. This latter problem should be partially alleviated if the survey is carried out by researchers external to the organisation. Additionally, surveys should be repeated over time to detect any changes effected by interventions to reduce bullying.

A survey that explores wider issues of staff work experiences and well-being may elicit more reliable results (e.g. Quine, 1999). Asking people about their attitudes and how they treat others may be as revealing as how they feel that they are treated themselves (e.g. Baron et al., 1999). Additionally, it is important to ask staff about witnessing others being submitted to inappropriate behaviour.

It is necessary to be aware that the self-report measures used in questionnaires suffer from a 'social desirability response bias', which means that people have a tendency to respond in a way that they perceive to be socially acceptable to those who will receive the information. In relation to interpersonal violence and aggression, people have a greater willingness to admit victimisation and a lowered tendency to admit perpetration with increasing severity of violence (Saunders, 1991). The questionnaire should gather information about attitudes, well-being, etc. before exposure to bullying behaviours is explored, then people should be asked about their experiences of different types of behaviour without, at this stage, using the term 'bullied'. Leaving the term until late in the questionnaire should reduce some effects of bias.

Staff appraisals

Appraisals can be designed to alert managers to interpersonal problems among their staff, particularly if the exercise is designed as a mutual learning exercise rather than simply as an assessment of a subordinate by a superior. Some organisations include 'upwards' and 'sideways' appraisal, so that employees are assessed by their colleagues and subordinates as well as by their superiors. Such systems might identify any bullying at an early stage as well as discourage unreasonable behaviour in the first place.

Objective evidence

Staff who are bullied have been shown to display relatively poor levels of job satisfaction, organisational commitment, intention to leave, physical health and mental health (Quine, 1999; Hoel and Cooper, 2000). It makes sense, with this kind of evidence, routinely to ask people leaving the organisation about any experience of bullying that might have contributed to their decision to leave. Those taking stress-related absences should be asked similar questions. Audit statistics should include bullying as a category wherever appropriate, such as for absenteeism and exit surveys, occupational health service reports and disciplinary proceedings. Any help-line, whether internal or external, should also be required to provide summary statistics including the incidence of problems caused by bullying and other forms of harassment.

Such objective evidence can provide warning signals to prompt further investigation. Such signs include higher than normal turnover or absenteeism in a particular department or team, a decrease in productivity of individuals or teams and a rapid turnover of new or young staff.

Discussion

National surveys indicate that the problem of bullying is widespread, affecting all sectors of industry to varying degrees (Hoel and Cooper, 2000). Thus, bullying is seen to be a serious issue that merits the attention of all organisations. Bullying can be tackled using the integrated organisational approach that has been endorsed for other health and safety issues including work-related violence (Leather et al., 1998). Such an approach requires the problem to be tackled at the levels of the organisation, the work team and the individual by the incorporation of risk reduction measures into policies, procedures, systems, practice and behaviour. These risk-reduction measures should address the problem at three stages: before, during and after incidents occur. These measures comprise prevention and preparation, timely reaction, and rehabilitation and learning.

Figure 6.1 The integrated organisational approach

These are all set within the organisational culture that prevails and attempt to affect that culture to provide a safer working environment.

Monitoring is a vital part of such an approach. It feeds into the system at a number of stages so that any measures introduced are based on good quality information about the problem as it exists within the particular organisation. It feeds into risk assessment, by providing an estimate of the amount of bullying that is occurring and by identifying groups of workers who are most at risk and situations that increase those risks. Analysis of the situations and the nature of the bullying can help to suggest measures at organisational, work team and individual levels to prevent recurrence. Monitoring can then be used to evaluate the effectiveness of the measures that have been implemented.

However, monitoring of bullying is seen to be a complex exercise that requires the provision of both formal and informal methods of reporting, with systems of support for those making a report, as well as the acquisition of information from a range of other sources such as staff surveys. Further, monitoring requires constant vigilance on the part of line managers, human resources managers, occupational health practitioners, and health and safety advisers. Catching problems at an early stage can prevent great detriment, not only to the individuals concerned but also to the organisation as a whole. The earlier that problems are detected and satisfactorily dealt with, the lower is the likelihood of bullying becoming an established part of the organisational culture.

References

Alderman, C. (1997) 'Bullying in the workplace: a survey', *Nursing Standard*, 11, 35, 22–4.

Andersson, L. M. and Pearson, C. M. (1999) 'Tit for tat?: the spiraling effect of incivility in the workplace', *Academy of Management Review*, 24, 3, 452–71.

Aquino, K., Grover, S. L., Bradfield, M. and Allen, D. G. (1999) 'The effects of negative affectivity, hierarchical status, and self-determination on workplace victimization', *Academy of Management Journal*, 42, 3, 260–72.

Ashforth, B. (1994) 'Petty tyranny in organisations', *Human Relations*, 47, 755–78.

Baron, R. A., Neuman, J. H. and Geddes, D. (1999) 'Social and personal determinants of workplace aggression: evidence for the impact of perceived injustice and the type A behavior pattern, *Aggressive Behavior*, 25, 4, 281–96.

Beale, D. (1999) 'Monitoring violent incidents' in P. Leather, C. Brady, C. Lawrence, D. Beale and T. Cox (eds) *Work-related Violence: Assessment and Intervention*, London: Routledge, 69–86.

Beale, D., Fletcher, B., Leather, P. and Cox, T. (1998) *Review on Violence to NHS Staff Working in the Community*, Nottingham: University of Nottingham.

Bies, R. J., Tripp, T. M. and Kramer, R. M. (1997) 'At the breaking point: cognitive and social dynamics of revenge in organizations' in R. A. Giacalone and J. Greenberg (eds) *Antisocial Behavior in Organizations*, Thousand Oaks, CA: Sage Publications, 18–36.

Bjorkqvist, K., Osterman, K. and Hjelt-Back, M. (1994) 'Aggression among university employees', *Aggressive Behavior*, 20, 173–84.

Bradfield, M. and Aquino, K. (1999) 'The effects of blame attributions and offender likableness on forgiveness and revenge in the workplace', *Journal of Management*, 25, 5, 607–31.

Cohen, D., Vandello, J., Puente, S. and Rantilla, A. (1999) 'When you call me that, smile!: How norms for politeness, interaction styles, and aggression work together in southern culture', *Social Psychology Quarterly*, 62, 3, 257–75.

Cooper, C. and Hoel, H. (2000) *Destructive Interpersonal Conflict and Bullying at Work: Key Findings*, Manchester: Manchester School of Management, UMIST.

Cox, T. and Cox, S. (1993) *Psychosocial and Organizational Hazards: Control and Monitoring in the Workplace*, European Occupational Health Series no. 5, Copenhagen: World Health Organization, Regional Office for Europe.

Durdle, T. (1997) *Report on the CPHVA Management Relations Survey*, Hereford: Durdle Davies.

Einarsen, S. (1999) 'The nature and causes of bullying at work', *International Journal of Manpower*, 20, 1–2, 16–27.

Einarsen, S., Raknes, B. I., Matthiesen, S. B. and Hellesøy, O. H. (1996) 'Bullying at work and its relationships with health complaints: moderating effects of social support and personality', *Nordisk Psykologi*, 48, 2, 116–37.

Employee Health Bulletin (1999) 'Bullying at work: a survey of 157 employers', *Employee Health Bulletin*, 8 April 1999, 4–20.

Farrell, G. A. (1999) 'Aggression in clinical settings: nurses' views: a follow-up study', *Journal of Advanced Nursing*, 29, 3, 532–41.

Folger, R. and Baron, R. A. (1996) 'Violence and hostility at work: a model of reactions to perceived injustice' in G. R. VandenBos and E. Q. Bulatao (eds) *Violence on the Job*, Washington, DC: American Psychological Association, 51–85.

Fox, J. A. and Levin, J. (1994) 'Firing back: the growing threat of workplace homicide', *Annals of the American Academy of Political and Social Science*, 536, 16–30.

Greenberg, L. and Barling, J. (1999) 'Predicting employee aggression against co-workers, subordinates and supervisors: the roles of person behaviors and perceived workplace factors', *Journal of Organizational Behavior*, 20, 897–913.

Hoel, H. and Cooper, C. L. (2000) 'Workplace bullying in Britain: some key findings from a study of 5,300 employees', *Employee Health Bulletin*, 14, 6–9.

Ishmael, A. (1999) *No Excuse: Beat Bullying at Work: Facilitator's Guide*, London: The Industrial Society.

Ishmael, A. and Alemoru, B. (1999) *Harassment, Bullying and Violence at Work*, London: The Industrial Society.

Leather, C., Brady, C., Lawrence, C., Beale, D. and Cox, T. (eds) (1999) *Work-related Violence: Assessment and Intervention*, London: Routledge.

Leather, P., Cox, T., Beale, D. and Fletcher, B. (1998) *Safer Working in the Community: A Guide for NHS Managers and Staff on Reducing the Risks from Violence and Aggression*, London: Royal College of Nursing and NHS Executive.

Leyden, G. (1999) 'Reducing violence to teachers in the workplace: learning to make schools safe' in P. Leather, C. Brady, C. Lawrence, D. Beale and T. Cox (eds) *Work-Related Violence: Assessment and Intervention*, London: Routledge, 145–65.

Leymann, H. (1992) *Frå Mobbning til Utslagning i Arbetslivet* (*From Bullying to Expulsion from Working Life*), Stockholm: Publica, cited in Einarsen, 1999.

McDonald, J. J. and Lees-Haley, P. R. (1996) 'Personality-disorders in the workplace: how they may contribute to claims of employment law violations', *Employee Relations Law Journal*, 22, 1, 57–81.

Quine, L. (1999) 'Workplace bullying in NHS community trust: staff questionnaire survey', *British Medical Journal*, 318, 7178, 228–32.

Ramage, R. (1996) 'Mobbing in the workplace', *New Law Journal*, 25 October 1996, 1, 538–9.

Rayner, C. (1997) *Unacceptable Behaviour: Workplace Bullying Survey*, London: UNISON.

Rayner, C. and Hoel, H. (1997) 'A summary review of literature relating to workplace bullying', *Journal of Community and Applied Social Psychology*, 7, 181–91.

Richman, J. A., Rospenda, K. M., Nawyn, S. J., Flaherty, J. A., Fendrich, M., Drum, M. L. and Johnson, T. P. (1999) 'Sexual harassment and generalized workplace abuse among university employees: prevalence and mental health correlates', *American Journal of Public Health*, 89, 3, 358–63.

Saunders, D. G. (1991) 'Procedures for adjusting self-reports of violence for social desirability bias', *Journal of Interpersonal Violence*, 6, 3, 336–44.

Skarlicki, D. P. and Folger, R. (1997) 'Retaliation in the workplace: The roles of distributive, procedural and interactional justice', *Journal of Applied Psychology*, 82, 3, 434–43.

VandenBos, G. R. and Bulatao, E. Q. (eds) (1996) *Violence on the Job*, Washington, DC: American Psychological Association.

Van Dick, R., Wagner, U. and Petzel, T. (1999) 'Arbeitsbelastung und gesundheitliche beschwerden von lehrerinnen und lehrern: einflüsse von kontrollüberzeugungen, mobbing und sozialer unterstützung' ('Occupational stress and well-being of schoolteachers: the impact of locus of control, mobbing and social support') *Psychologie in Erziehung und Unterricht*, 46, 4, 269–80.

Warchol, G. (1998) *Workplace Violence, 1992–1996*, Report NCJ 168634, US. Department of Justice, Bureau of Justice Statistics.

Wolfgang, J. (1999) 'Mobbing in organizations: a review of the state of the art of mobbing research', *Zeitschrift fur Arbeits-und Organisationspsychologie*, 43, 1, 1–25.

Zapf, D. (1999) 'Organisational, work group related and personal causes of mobbing bullying at work', *International Journal of Manpower*, 20, 1–2, 70–85.

PART THREE

Imperative to act

CHAPTER SEVEN

Dignity at work

The legal framework

PATRICIA LEIGHTON

Introduction

This chapter covers two main issues. The first is an assessment of the nature, traditions, role and effectiveness of the law in responding to workplace bullying. The second is a critique of the law and the legal process itself and a questioning of whether the law is a help or hindrance to individuals and organisations when dealing with bullying.

It is important to assess the role of law in responding to bullying, not least because little is known about the impact of the law in changing behaviour in the workplace (despite the increasing volume of legal regulations relating to employment) and, more widely, about people's attitudes to the law. The natural assumption in Britain and in other parts of the EU is that bullying is a workplace problem. Bullying is seen to be detrimental to an individual employee and this should prompt a legal remedy for that individual. Law is thus both individualised and individual-remedy driven, despite the mounting evidence presented in this book that the causes of bullying, as well as the effective interventions to respond to it, are organisation-driven and complex.

The second section of this chapter explores the workings of the legal system, the role of lawyers and others in litigation and questions the validity of the law in its responses to bullying. This is especially important given the adversarial nature of litigation and the continuing importance of the concepts of 'liability' and 'fault'. The section goes on to explore whether recent and anticipated legal developments are appropriate and constructive or whether they will have the effect of fuelling a litigation culture.

There are some subtle or even entirely unarticulated factors in all legal systems that can produce a backlash or be undermining when people (including those on the Clapham omnibus) consider the law has gone too far. The supporting of causes or individuals regarded as unworthy, or by making everyday life and decision-making over-difficult or complicated can cause such reactions from the general public. These common views sometimes appear in the tabloid press, in letters to 'broadsheets', in parliament and in professional literature; less so in academic works. However the power of these views should not be underestimated. Silent acquiescence, if not open support by ordinary working people is critical for successful litigation and case law. The politics of the law and the legal process have especial relevance to a topic such as bullying that is both hard to define or explain and is possibly controversial.

The last part of the chapter examines the traditions of British law and legal processes and describes changes to the procedural aspects of courts and tribunals. A further issue that is touched on is the role of discretion in initiating legal action, especially the enforcement of statutory duties. In many debates about law an often-neglected dimension is the analysis of when, why and by whom steps are taken to enforce duties. Particular prominence is given to the health and safety inspectors and environmental health inspectors who have responsibility for enforcing health and safety legislation. Finally there is an examination of the factors that appear to influence the individual when seeking redress through the courts for being bullied.

The legal framework

The immediate curiosity of the legal framework is, potentially, how *so much* law has relevance for bullying. However, the first critical issue is the question of definition.

The law and lawyers traditionally delight in tussling with and then providing definitions which are both accurate and comprehensive. The law is more comfortable in providing definitions where the subject matter is tangible and, if harm is involved, where that harm is medically recognisable. Such words as 'harassment', 'victimisation', 'bullying', 'threats' and 'violence' are all at different times applied to circumstances where the conduct involved is similar. Indeed the factor which often causes an individual to resort to law or that causes the law enforcement authorities to intervene, is stress or other psychological illness which can be the consequence of all such behaviour. It needs to be recognised that the traditional legal approach to workplace issues is first to focus on the harm or detriment such as a job loss, industrial accident or unequal pay. The aim then is to provide a remedy on the basis of the employer's breach of a statutory or common law duty. The concept of rights at work, coupled with an entitlement to a remedy where rights are infringed, is a relatively recent legal construct. Hence the idea of an entitlement to being treated with dignity at work, which has to be shown by employers and colleagues, let alone customers and clients, is only gradually being absorbed. The EC recommendation, 'Dignity at work' (1991), which spearheaded this new legal thinking has only slowly had impact on both employers and UK law. This is an important change because a 'rights' approach inevitably focuses on the individual victim and their expectations, in contrast to the traditional legal approach which examines a defendant's alleged breach of statute or common law.

The current legal framework of relevance to bullying includes both these approaches and, in principle, opens up a range of legal remedies. In theory, this richness of laws should ensure that victims of bullying are able to obtain redress for their suffering.

Definitions

It has already been mentioned that, from a legal perspective, bullying appears hard to define or at least to contrast with other forms of behaviour such as harassment and victimisation. Lawyers tend to resist the use of 'popular' definitions or ones that are too jargon-ridden. For example, they have only recently come to terms with post-traumatic stress disorder, preferring the more accessible phrase 'nervous shock' as applying to victims of intentional or negligent behaviour subjecting them to psychological harm.

Some writers have included bullying and harassment as subsets of violence (Knorz and Zapf, 1996); others see harassment as racially- or gender-based where the essential factor is the individual's membership of a group (O'Donnell, 1999). Most consider bullying as a consequence of particular interpersonal or organisational factors: bullying is 'personal'.

In principle, the individualised aspect of bullying makes it easier to accommodate within traditional common law approaches. In contrast, race and sex discrimination laws, which examine behaviour in the context of whether the treatment of a member of a group had been detrimental if compared with the treatment of a non-member of that group, has produced highly complex and often unsatisfactory case law. Much of the literature on discrimination law, including new discrimination areas that continue with this 'comparator' approach, is highly critical, pointing out that improvements in the position of those in minority groups have been slow and only marginal (Ashiagbor, 1999). Many highlight this 'comparator' approach as being a major factor in the lack of progress.

Bullying has attracted a variety of definitions, but most writers simply focus on examples of bullying behaviour and the effect on the victim. Widely accepted instances of bullying include unfair and excessive criticism, insults in public, constantly ignoring or excluding the victim from discussions, decision-making or social activity at work, undermining the victim, devaluing their role and work and shouting or sending threatening memos, e-mails, etc. Some forms of bullying are open and easily observable; others are more subtle and pervasive.

With a growing emphasis on understanding the causes of bullying there has emerged a clearer picture of it 'in the round'. The perpetrators appear just as likely to be women as men, though bullies are more likely to be in a position of authority. Occasionally the perpetrator is a subordinate. There is also some evidence of group bullying, especially by colleagues (Crawford, 1992).

The typical characteristics of bullies appear to be that they themselves have been victimised by bullying, they tend to be divisive or disruptive at work, are bereft of empathy, blame others for mistakes, resist apologising, cannot distinguish between assertiveness and aggression, often have mood swings and are uncommunicative. Men tend to be more overtly hostile to victims; women are more manipulative and secretive in their bullying (Mantel, 1994).

Research undertaken for the Trades Union Congress has highlighted the relevance of the organisational or situational context to an understanding of the causes of bullying (*Employee Health Bulletin*, 1999). Its incidence in the public sector has been especially widely researched and analysed (UNISON, 1997). Under-resourcing and a growing volume of work and stressful situations, especially in emergency services, appear to increase pressure on managers such that the reported incidence of bullying is high. Social work, the education service and government departments that have considerable interface with members of the public appear especially susceptible.

The consequences for victims are similarly well researched (Leymann, 1992; Niedl, 1996). Victims report illness, often leading to resignation from their work following 'sustained campaigns against them'. Commonly the victims will describe 'life being made a hell' and being unable to 'face coming to work' (Field, 1996). However, clearly people do react differently. A bullying behaviour might cause one victim to complain, invoke the grievance procedure or confront the bully, another victim of the same bullying behaviour might have a complete breakdown or, in extreme cases, take their own life. The consequences of workplace bullying are unpredictable and highly personalised, and are often influenced by other factors in the victim's life, such as bereavement, relationship breakdown or other anxieties unconnected with work.

Less well researched, but also important for a legal analysis, is the question of the personality and professional abilities of the alleged bully (Gandolfo, 1995). From a senior management standpoint the alleged bully might be charismatic, a strong motivator and highly effective, especially in dealing with inefficiency and laziness by staff. There may be considerable perceived benefits to the employing organisations; it may be recognised that the manager in question has a challenging approach and needs to make difficult decisions. Indeed, the organisation itself may thereby encourage bullying behaviour or, at least, turn a blind eye to its incidence. Senior managers may themselves be characterised as aggressive or even as bullies.

Alternatively, the organisation may be totally unaware of bullying behaviour by managers or staff. The organisation may consider high levels of sickness absence or labour turnover as typical of their industry or even an indicator of a fast-moving, lively and successful market leader. Where the bullying takes a more subtle form, their ignorance may be totally genuine. From a legal standpoint all the above factors and data drawn from research have considerable relevance.

As will be seen, there are many areas of law which can be applied to bullying, each having distinct approaches and rules. Some areas require clear evidence of physiological or psychological harm to the victim and actual awareness by the employer of what was going on. For other areas of law a court will state that they ought to have known that bullying was occurring even if they deny actual knowledge.

There is a need to consider the question of the response of the victim to being bullied. Some laws, such as the law of negligence, require a degree of robustness from alleged victims. A reaction to, say, pointed criticism, of suffering depression and being on sick leave for several months might be considered excessive, especially where the victim has a medical record of vulnerability or psychological illness. The law here has taken the view that 'reasonable' people should be able to cope with the hurly burly of modern life (Robertson vs Forth Bridge Company, 1995).

Individual employees, even senior managers, are rarely worth suing for negligence; the question then arises of the vicarious liability of the employer for employee acts of bullying. Specifically, how relevant is it whether the employer knew or might have known of the bullying? Perhaps they had considered that there might be a personal vendetta by the manager which was only loosely connected with work.

Other areas of law, such as the statutory health and safety framework, focus on the legal demands on the employer to comply with legal rules in order to ensure health and welfare of workers. The fact there has been no victim or illness does not affect the legal position regarding a breach of the relevant regulations. In these circumstances a close examination of the employer's policies and practices is undertaken by the health and safety inspectorates to ensure that the organisation has measures in place, and applied, to prevent harm through bullying. The fact that there have been instances of illness, absence and civil claims in negligence might have alerted the inspectorates but would not be technically relevant to their actions.

However, in other areas of law, for example those involving the workings of the contract of employment, including the application of disciplinary rules and dismissal, there is a requirement to examine the impact on the alleged victim *and* the behaviour of the alleged bully, all set in the organisational context. Most discrimination legislation adopts the same broad, holistic approach, though the application of detailed statutory rules is distinctive. The motives of the alleged bully may or may not have relevance, depending on the area of law. Similarly, it may or may not be necessary to compare the behaviour of an individual or an organisation with a hypothetical reasonable and competent employer in a similar position to the defendant in question.

To sum up: different areas of law adopt different and distinctive approaches to bullying as the basis of legal action. The various areas of law examine the factual situation differently and have different expectations of the parties involved. Rules have emerged through different legal traditions. These include the common law, in particular the law relating to trespass to the person (assault and battery), negligence and the law of contract. There is then the statutory employment framework, now increasingly influenced by EU law, as well as legislation relating to rights at work, including job security and anti-discrimination legislation. Finally, there is a growing body of law, in particular the 1991 European recommendation on 'Dignity at work', which is entirely 'Euro-driven'. These, too, have a very different legal tradition and one that is rooted in the civil (Roman) law tradition of all EU member states.

Constraints on legal actions

There is another important legal issue which underpins virtually all potential legal actions brought by those suffering from bullying. This is the so-called 'floodgates' argument. The importance of this argument should never be underestimated. It represents a control mechanism with which the legal system can prevent courts and tribunals from becoming overwhelmed by particular types of claims. There is evidence that since the first stress case – Walker vs Northumberland CC (1995) – it has proved very difficult for claimants to succeed with the result that negligence cases are proving difficult to win.

It is also important to reflect on the fact that, for many lawyers and courts, litigation should always be the last resort. There is a firmly held belief that 'local' organisational procedures should be exhausted before there is recourse to formal legal action. Indeed, as is illustrated below, recent changes to the rules of civil procedure have tended to emphasise this fact. Cases that lack substance, which are susceptible to arbitration, mediation or conciliation, or where there are doubts as to the genuineness of the claim can be now struck out or referred back. The legal system has signalled the fact that it is only well-reasoned, well-documented and legally-robust claims that should go forward. It is inevitable that claims for bullying, especially where the evidence is flimsy or controverted, might well be more difficult to sustain in the future. At the same time, providing there is a clear articulation and a growing consensus on the essential nature of bullying and evidence to support an individual claim, case law should continue to evolve to provide remedies.

The working of the legal system and role of lawyers

Legal remedies

It will be argued later that there are some legal processes, especially those relying on the contract of employment, which appear to offer the most appropriate, effective and comprehensive approach to bullying at work. However there are other legal rules, mostly well established and considered, which have obvious relevance to bullying.

Legal rules are categorised in the following way:

1 Those which are victim driven, i.e. those that focus on the impact on the physical or psychological well-being of an individual and where the law aims to provide appropriate compensation to the victim for this detriment. The

areas of law of particular relevance are: discrimination, negligence, victimisation under the Employment Rights Act (1996), and human rights law. All these areas of law can raise issues about organisational responsibility.

2 Those which are allegedly 'bully' driven, i.e. those that explore the conduct of an individual, typically a manager. These legal rules focus on the intentional behaviour of the alleged bully and question the extent to which the employing organisation is responsible for it. The law of trespass to the person appears to have an emerging relevance, though, as yet, the case law is very limited. It is important to recognise the role of criminal law applying to bullying. It may be that the Protection from Harassment Act of 1997 has relevance. It is also important to note the possibility of the conduct of the bully leading to awards of compensation for victims as a result of breaches of anti-discrimination legislation which also creates 'statutory torts'. This legal mechanism provides for compensation for victims of statutory breaches. The other important area of criminal law is the health and safety statutory framework. This can focus on both the individual and the employing organisation and has a number of specific legal rules and requirements which can apply to bullying.
3 Those areas of law which adopt more of an *overview approach*, i.e. those that examine both the behaviour and harm suffered by the victim and the conduct of individuals within an employing organisation and the organisation itself.

The role of the contract of employment and, in particular, the implied terms of co-operation, and mutual trust and confidence will be explored, along with current approaches to using the employment contract as a vehicle for obtaining redress for being bullied. The EC recommendation, 'Dignity at work' (1991) and the proposals to create an explicit piece of UK legislation will also be touched on.

With the possible exception of the Dignity at Work Bill (1999), none of the legal rules referred to above is explicitly and specifically designed to provide protection from bullying. It is necessary to examine the extent to which broadly-based areas of law can be legitimately applied to bullying. The case law is thus neither simple nor always convincing. It is perhaps inevitable that behaviour, which many professional groups find difficult to define and explain, also presents challenges to the law and to lawyers!

Victim-driven areas of law

Anti-discrimination law

This was probably the first area of statute law to address bullying. A claimant has to prove that they have suffered a detriment such as an illness, job loss, reduced promotion opportunities or job satisfaction. They have then to identify the cause, in this case, bullying, whereby the conduct complained of related to the claimant's sex, race or disability. Ironically, in an organisation where bullying is rife and indiscriminate it can prove almost impossible to bring a case relying on the Sex Discrimination Act (1975) or Race Relations Act (1976). There is no comparator to be found of another sex or race that has been treated better!

Nonetheless, case law illustrates some extreme conduct which, as well as being discriminatory, amounted to assaults. In Bracebridge vs Darby (1990) a woman was abused and assaulted on a billiard table by her supervisor; in Clayton vs Hereford and Worcester Fire Service (1997, unreported) humiliating and sexual initiation rituals gave rise to an award of £200,000 compensation, an exceptional amount.

To sum up: anti-discrimination legislation has not always proved especially helpful in bullying situations. The onus on the victim to produce evidence that it was *because* of their sex or race that they were subjected to bullying is not always easy to discharge. Today it is unusual for bullies to say such things as 'I'll get you, you black/bastard/tart/bitch'. As the composition of the labour market reveals an ever-growing number of women and people from ethnic groups, it can only get more difficult to establish the gender or racial basis as the root cause of the bullying. Where the bullying appears as a personality clash, envy, a wish to undermine or deal with a perceived threat, it will not easily be seen as racially or sexually motivated. Furthermore, assuming a successful anti-discrimination claim, there are other problems. Firstly, the question of remedies. Victims, typically, want the bullying to end, the perpetrator removed and their working life to be harmonious. However, UK legal remedies tend to be 'in rem.', that is they offer financial compensation, rather than 'in personam', such as injunctions or specific performance. Courts and lawyers usually see the initiation of litigation as damaging to the essentially interdependent relationship of employer and employee. There is, therefore, considerable reluctance to provide remedies other than in a cash form and these are typically awarded after the claimant has left the employment. In any case, sex and race discrimination litigation tends to be particularly fraught and contentious.

The second problem is related to the first: this is who will pay? Levels of anti-discrimination compensation are rising so the insurance company of the employer is often the only viable source of compensation. To make the employer liable for the discriminatory conduct, it must be directly or vicariously liable. The former implies that senior managers encouraged or participated in the bullying or harassment, the latter that the employer was responsible for the behaviour of others including, perhaps, non-employees.

Perhaps the most notorious case here is Burton and Rhule vs De Vere Hotels (1997) where Bernard Manning subjected two black waitresses to racist abuse during a performance at a hotel. The employer was held liable under the Race Relations Act (1976). The hotel management was found to have known, or deliberately or recklessly closed its eyes to what was happening and failed to do anything to prevent it. Its liability depended on the organisation's ability to control events. This approach can be helpful to victims in the case of bullying in that it emphasises that proclaimed ignorance of what is happening does not inevitably provide employers with immunity.

Can there be vicarious liability? Is the employer liable for discriminatory conduct by managers or other employees? This has proved a surprisingly contentious issue because tribunals have sometimes concluded that bullying behaviour is outside the scope of employment and thus the employer will not be vicariously liable. In effect, the employee is not employed to bully; they are there to do their work. Indeed, the irony is that the more outrageous the bullying, the more some tribunals think it is likely to be outside the scope of employment. There are now some signs that this view is changing. Arguably it must, as to find otherwise allows employers to condone or at least turn a blind eye to bullying. In Jones vs Tower Boot Company (1997) the black complainant suffered the most dreadful abuse and bullying by colleagues. This included serious physical assaults. The employers were held liable, despite the nature of the conduct, as the employees concerned were considered to be 'in the scope of employment'. They were on work premises, during working time and their employer was vicariously liable for them.

Overall, anti-discrimination law has proved helpful to provide remedies for some forms of bullying and racial abuse.

Discrimination law and compensation for psychiatric injury

A new and potentially highly significant development arises from the Court of Appeal's decision in Sherrif vs Klyne Tugs Ltd (1999). Sherrif, a Somali muslim was employed as an engineer on a tug. He was harassed and bullied by the master of the tug. He settled the claim of £4,000 for race discrimination following ill-health termination of his contract. He then claimed in the County Court for compensation for personal injury (a nervous breakdown). The County Court stated that the claim had been settled within the £4,000. The claimant appealed to the Court of Appeal. This confirmed that compensation for personal injuries could be obtained in a tribunal as a statutory tort and as an addition to injury to feelings.

There is much current debate as to the significance of this decision, not least that claiming personal injury compensation might be easier through a statutory tort derived from anti-discrimination legislation than through the law of negligence. The statutory tort does not require evidence of reasonable foresight of harm on the part of the employer so as to establish the duty of care. Many 'vulnerable' victims of bullying have lost cases on the basis that the law only requires reasonable foresight of normally robust people.

Even though there are doubts about the exact scope of this statutory tort, it appears the decision does open up new possibilities for compensation for some forms of harassment and bullying.

The law of negligence

This area of law is flexible, evolving yet well established. It focuses on the need for reasonable care to be shown to victims of physical or psychological harm. In principle, some victims of bullying should be able to rely on it to establish that, by continuing to employ a bully and not curb their bullying behaviour, this is a breach of the duty of care owed to all employees. As considered above, the legal reality has been that successful claims are few and far between, not least because the need for the victim to establish a reasonable foresight of harm by the employer on the specific facts of the case can be easily thwarted. This is because the law only requires reasonable foresight of harm in a normally robust person. Hence, vulnerability, emotional instability, depressive illnesses and similar states can lead a court to conclude that this was the reason that the individual became a victim and that the conduct to which they were exposed was within the range which a 'normal' person could have withstood. Ironically it is this very vulnerability which makes some victims an easy target for bullies. The failure of some victims to overcome this legal rule, especially coupled with the traditional reluctance of courts to open the 'floodgates' to litigation, is an important issue. Even where 'whistle-blowing' has activated the bullying, the courts remain resistant to negligence claims.

The law of negligence has other intrinsic problems – does the employer have to have *actual* knowledge of the bullying and the harm? It is not absolutely clear in law whether this is the case. It is easier for a victim to win if the organisation is aware of the problems via a grievance, memo or formal complaint. There is also the question of the 'standard of care' which should be shown to the victim of bullying. The employer must show 'reasonable' care (the degree of care which a reasonable employer would display). This reasonable care, in turn, should reflect the established practices in employing organisations at the relevant time. The standard need not be higher, in reality, the standard is what a judge thinks is a correct standard – not too demanding or protective of 'wimps', nor too protective of an organisation that let bullies run riot!

The law of negligence has not yet provided much help to victims, largely owing to the conservative and incremental way the common law evolves in Britain. It is also intrinsically problematic to try to find a remedy in a cause of action which is based on careless, neglectful behaviour rather than intentional. Bullying is essentially a direct and intentional type of behaviour.

Trespass to the person: assault

This is perhaps the most interesting and potentially most relevant area of law. The essence of law is that the conduct of the defendant causes the claimant to be fearful, apprehensive or otherwise concerned for his or her welfare. There is no necessity for physical or psychological harm. The law's concentration is the effect of the bullying conduct – especially threats, abuse and the more obvious and direct forms for bullying – on the victim. It is thought that apprehension of psychological harm could suffice. Early case law dating from the nineteenth century had established that intentional conduct causing psychological harm could lead to successful claims (Wilkinson vs Downton, 1897). Arguments that it was only a joke, that people over-reacted or conduct had been misinterpreted have only rarely been successful.

In 1998 an interesting Scottish case – Ward vs Scotrail (1998) – a classic bullying and harassment case emerged. A railway worker was subjected to sustained intimidatory behaviour from a colleague. She became ill and her employers failed to provide her with support or deal with the bully. In a very interesting case, the judges tussled with the various legal rules which could potentially apply. The role of the law of trespass was accepted as an appropriate course of action and there was much discussion on possible vicarious liability.

For this area of law to be really helpful to victims of bullying, or at least some forms of bullying, vicarious liability is a critical issue. This hinges on whether the trespass was in the 'course of employment', an issue previously considered. However if this matter can be resolved, the law of trespass is simple and direct. It does not require job loss or comparisons between treatment of one sex or race and another and the motives of the defendant are especially important.

Areas of law which focus on the bully or the bully's employer

Both criminal and civil law have relevance. Clearly, some forms of bullying can amount to crimes of assault, such as threatening behaviour and indecent assault, and the perpetrator can be punished.

The Prevention of Harassment Act (1997) – the 'stalker' legislation – provides criminal and civil sanctions and can potentially apply to workplace bullying. The act aims to curb persistent intimidatory conduct and some lawyers see a role for a discrete and comparable piece of legislation applying to the workplace. Under this legislation, significant fines or even imprisonment can be imposed. Compensation can also be ordered for the victim on the basis of a statutory tort.

Criminal law normally concerns itself with individual misconduct. Many researchers and writers on bullying see collective bullying as important as individual hectoring or undermining behaviour. The area of criminal law which appears to have most relevance to collective or organisation behaviours is the health and safety legislative framework. This is enforced through criminal sanctions.

The Management of Health and Safety at Work Regulations (1999), implementing the EC Framework Directive (1989), on health and safety, requires managers in all organisations to ensure the health, safety and welfare of employees and others. In particular they are required to undertake risk assessments of all potential workplace hazards, implement appropriate preventive measures to deal with matters and monitor the effectiveness of those measures. Where the hazards are fire, dust, chemicals or even physical violence, a risk assessment is required. When the risk is bullying, the task is clearly more difficult and highly sensitive. Indeed, the closer the examination of the legislative framework, the less likely it appears to be able to provide a ready solution to deal with bullying.

The health and safety inspectorates are notoriously unwilling to intervene or prosecute in these controversial or more imprecise areas. The health and safety regulations also do not provide for a civil claim for compensation.

The only circumstances where it is likely that health and safety legislation will be applied is where an organisation's tolerance of a bullying culture has led to a high incidence of illnesses, especially stress-related disorders. It is possible that, as an alternative to prosecution, the inspectorates would impose an improvement notice requiring an effective anti-bullying strategy and practices.

The 'overview' legal approaches

A rapidly growing area of interest, research and analysis is the role of the contract of employment; in particular the implied term of trust and support which is present in all employment contracts. This requires the employer to be aware of workplace problems, to respond to complaints and generally to ensure measures are in place to provide employee well-being. Clearly, the failure to deal with bullying and provide support for victims potentially breaches this term. Breaches then *can* lead to claims for constructive dismissal and unfair dismissal and there is a rich and long-standing body of case law illustrating the operation of this contractual rule to a wide range of circumstances.

The approach of employment tribunals is to explore the work situation in the round, to apply the tribunal members' own knowledge and experience of the world of work to the facts and to make assessment based on reasonableness and custom and practice. In this process the conduct of the bully, the victim and the employing organisation itself has relevance and is weighed carefully. The characteristics of the employment sector, of its workforce and of particular circumstances, which might account for the bullying or victim response, can therefore all be examined.

Some cases illustrate the application of the implied term of trust and support to typical workplace bullying situations: In Hardman vs Brake Brothers Food Services Ltd, an employee was regularly criticised about work performance and told she had 'an attitude problem' and behaved 'like a child'. In reality, her work was not good and the tribunal confirmed that employees must accept reprimands. However (and this is the key) the way in which criticism was delivered to the employee was excessive and therefore a breach of the implied term. On the other hand, the implied term was held not to have been broken in Bowers vs Londis (Holdings) Ltd where an employee alleged that the arrival of a new accountant made her work 'unbearable' and that he was rude to her. However, in the verbal exchanges between the two, the claimant was apparently 'capable of giving as good as she got'. There may well have been an issue of inadequate supervision or management in the department but there was no evidence of breach of the implied term in the facts.

A particularly interesting and, from a practical point of view, difficult case is that of Adams vs Southampton and SW Hants Health Authority. An employee refused to participate in industrial action. As a result, her colleagues 'sent her to Coventry', she received threatening phone calls and her locker was broken into. She won her case, as her employer should have realised she needed support from this type of bullying. However, from an employer perspective, this poses a tricky problem. Protecting the employee might have prompted further industrial action; other than offering her a severance package or moving her to an entirely different part of the organisation (which may not have been feasible or may not have served to protect her), dealing with the problem is difficult. It is inevitable that bullying by a colleague or colleagues (who could be *individually* sued or even prosecuted) presents major dilemmas.

Although cases brought in employment tribunals for either unfair dismissal or breach of contract, relying on the implied term, are unpredictable they are likely to be handled in a practical, employment-orientated manner.

It must also be borne in mind that the employment contract of all employees contains an implied entitlement to a safe and healthy workplace. An employer who encourages or condones bullying can, potentially, also be considered to be in breach of this term of contract. As yet, case law directly on bullying or harassment has not evolved in this area.

Overall, the law already provides a richness of potential legal remedies, though they all have shortcomings. The next area to turn to is the legal process itself.

Traditions of law and legal processes

The legal process and bullying

Although this point will be considered more fully later, it has to be borne in mind that taking legal action in itself is an aggressive act. It will question, challenge, seek to undermine, highlight friction at work and be generally an unedifying, divisive and stressful activity.

A genuine victim of bullying, especially one who has suffered severe health problems, will have to weigh carefully the potential benefits to be gained through a successful claim against the stress, time and longer-term consequences of litigation.

The following questions will need to be faced by a potential litigant:

Can I afford litigation?

There is no legal aid for employment tribunals and legal aid for personal injury cases (e.g. negligence and trespass) has been recently severely curtailed. Contingency fee agreements (CFAs) are available, whereby the lawyer 'takes a cut' of the damages, however is not paid if the case is lost, but lawyers are not naturally keen to take 'test' or speculative cases where the chance of success is low.

Trade unions may support a claim, though where the alleged bullying is by colleagues or they consider that litigation will sour industrial relations, they are sometimes notoriously reluctant to do so. Anecdotal evidence suggests that some unions are generally more supportive than others. Without legal aid, or the financial support of a trade union, a litigant has to consider the financial implications, not just should they win, but should they lose. They can be dire.

It is also the case that the option of saving costs by being a litigant in person is neither a particularly easy one for the litigant nor typically successful. Courts, rightly or wrongly, consider that such litigants often delay and confuse the legal process.

Do I have the evidence?

This question has two elements. First, can I easily gain access to the evidence, say, of an internal memo stating that you are a 'nuisance' and need to be 'dealt with' or 'taught a lesson'? Are there witnesses? Is there medical evidence – and how robust is it?

Second, will the evidence stand up? Will work colleagues really support you against management? Will your response simply look like hypersensitivity or truculence? Is it likely that the employer's witnesses will be more reliable? Will they stick together?

Can I sustain the process?

Even allowing for the Woolf Reforms that aim to speed up and simplify civil litigation, in essence the process is still unlikely to be quick or easy. The law in Britain remains an adversarial process, geared to producing a 'winner' and 'loser'. There is the possibility of the Advisory, Conciliation and Arbitration Service (ACAS) helping to resolve a tribunal claim but the chances of arbitration, or an alternative dispute resolution (ADR) through mediation, remain a remote possibility. The litigation is geared to a 'fight to the finish' with all that that entails.

Can I handle the hearing?

Hearings also follow the adversarial style with witness examination and cross-examination; the latter usually geared to undermining the evidence of the claimant or witness. Hearings are tightly structured and follow a traditional process. Unfamiliarity or resistance to this will inevitably cause a claimant, especially a litigant in person, considerable problems. In bullying cases, character, motives and previous behaviour are at the heart of the hearing. Hearings can be very acrimonious and demeaning.

Will I get the remedy I want?

First, of course, what are the chances of success? It remains the situation that for bullying cases the success rate remains low. Even if a claimant *is* successful in some areas of law, such as in a case of negligence and unfair dismissal, the 'contributory' conduct of the claimant will be taken into account. If compensation is awarded, it might then be reduced accordingly.

Courts and tribunals are unwilling to consider remedies other than compensation as appropriate as a bullying case almost invariably indicates a breakdown in work relations. Victims will need to seek other employment and will need good references from their ex-employer. Where a settlement is negotiated, this is often an important element.

Tribunals and courts do not have the ability to, say, order the disciplining or dismissal of the individual or group of bullies. On the whole, therefore, the victim is likely to receive a cash payment and coverage of their costs if they are successful.

The well-informed 'victim' will have to take on board these important issues.

The law relating to bullying's impact on employers

In truth, we still know very little about the precise extent to which employers react to legal rules in terms of making practical changes in their organisation. It would be comforting to think that a headline case, say on stress caused by bullying, triggered a major senior management review, followed by the introduction and enforcement of a series of policy initiatives to deal with any relevant problems. This does not appear to be what happens. This is not to say that change does not take place, but it is often more of a procedural, defensive nature than one affecting the whole culture and workings of an organisation.

Dealing effectively with bullying may require a careful and difficult assessment of the mission, goals, objectives, management staff, training, promotion criteria and practices, the role of human resource management and, generally, the image of the organisation. On the other hand, it may simply require more effective supervision, complaints and support processes, and general awareness raising.

Reflecting on the range and nature of legislation should lead employing organisations carefully to consider that bullying has a legal as well as organisational realism.

All of the following reactions appear to be misconceived:

- 'That ours is an organisation where bullying simply does not happen – there is no need to make enquiries or take steps.'
- 'Maybe bullying does go on, a bit, but surely no employee is going to be so stupid as to sue?'
- 'Even if someone sues they are not likely to win. Bullying is such a vague matter; the law has more important things to deal with.'
- 'Any sensible court or tribunal is bound to see X as a trouble-maker or deranged. They will understand we have a business to run and all we were doing was pursuing our need to be efficient.'

- 'There are not going to be new laws on bullying – surely the experience of trying to legislate for dangerous dogs, for guns and knives shows how futile such laws can be?'

More typical organisational responses have been to see bullying as a genuine workplace issue and to recognise that taking positive steps to deal with it makes legal as well as organisational sense. In a court or tribunal it is an obvious error to question whether bullying exists or why a policy to deal with it might be required.

At the least, organisations need the following:

- Awareness and acceptance, especially by senior management, that bullying might be taking place.
- An overt and continuing commitment to dignity at work for all people and a 'zero' tolerance of bullying.
- Appropriate policy development, perhaps via health and safety policies, to deal with bullying. There might be a cross-reference to an equal opportunities policy, but this is not essential.
- Appropriate procedures and support for those who consider themselves victims, including confidentiality, if required. Confidentiality may make it more difficult to deal with the problem. The issues relating to confidentiality need to be weighed carefully in the context of preventing further bullying.
- Training and support for managers/supervisors, in particular how to handle difficult situations sensitively, with the avoidance of bullying.
- A strategy to deal with those accused of bullying regarding the immediate allegations and their future work.
- A recognition that bullying is a complex problem, not susceptible to quick-fire solutions. Courts and tribunals appear to appreciate this.

Issues and future directions in law

Bullying as a workplace problem does not lack either a legal framework or potential remedies. Indeed, the situation is complex and requires careful analysis and understanding. Also requiring careful analysis is the question of whether victims should resort to litigation and whether law itself is an appropriate avenue for aiming to reduce the incidence of bullying at work.

Our litigation process remains adversarial, as compared with the inquisitional process used in continental European jurisdictions, the latter providing a more 'paper-based' and investigative approach. Clearly, the position of litigants and witnesses in the British system, which has an oral and combative tradition, is stressful. The litigation can take a considerable length of time and be expensive.

Are there improvements taking place? To some extent, the answer is 'yes'. The so-called Woolf reforms aimed at simplifying and speeding the civil justice system may succeed. The reforms aim to screen out unwarranted cases and encourage early settlements. This mirrors changes in the employment tribunals from 1996, whereby litigants are encouraged to make 'compromise agreements' to settle cases, often with the support of ACAS. Indeed, employment litigation has long benefited from ACAS's conciliation work.

Outside tribunals, there has been much talk but relatively little action on alternative dispute resolution (ADR). Arbitration, mediation and conciliation happen but there seems relatively little enthusiasm. However, the reduction or withdrawal of legal aid concentrates the minds of litigants, and perhaps ADR will become more popular.

If there is change afoot will victims be better aided and, perhaps more importantly, will such changes lead to employing organisations in developing appropriate anti-bullying strategies? It is not clear, not least because we know little as yet of how law impacts generally on organisations.

References

Adams vs Southampton and SW Hants Health Authority (COIT 1560/156).

Ashiagbor, D. (1999) 'The intersection between gender and "race" in the labour market: lessons for anti-discrimination law' in A. Morriss and T. O'Donnell (eds) *Feminist Perspectives on Employment Law*, Cavendish Publishing.

Bowers vs Londis (Holdings) Ltd (COIT 3083/245).

Bracebridge vs Derby (1990) IRLR 289.

Burton and Rhule vs De Vere Hotels (1997).

Clayton vs Hereford and Worcester Fire Service (1997).

Crawford, N. (1992) 'The psychology of the bully' in *Bullying at Work: How to Confront It*, Andrea Adams, London: Virago Press.

Employee Health Bulletin (1999) 'Bullying at work: a survey of 157 employers', *Employee Health Bulletin*, 677, 4–20.

Field, T. (1996) *Bully In Sight: How to Predict, Resist, Challenge and Combat Workplace Bullying*, Wantage: Success Unlimited.

Gandolfo, R. (1995) 'MMPI–2 profiles of workers' compensation claimants who present with complaints of harassment', *Journal of Clinical Psychology*, 51, 711–15.

Hardman vs Brake Brothers Food Services Ltd (COIT 3026/122).

Jones vs Tower Boot Co. (1997) IRLR 168.

Knorz, C. and Zapf, D. (1996) 'Mobbing: a severe form of social stressor at work', Zeitschrift fur Arbeits und Organisationspsychologie, 40, 12–21.

Leymann, H. (1992) 'Psyiatriska problem vid vuxenmobbing: en rikstackande undersorning med', 2, 438, intervjuer, Delrapport 3, Stockholm: Arbetarskydd-styrelsen.

Management of Health and Safety at Work Act Regulations (1999).

Mantel, M. (1994) *Ticking Bombs: Diffusing Violence in the Workplace*, New York: Irwin.

Niedl, K. (1996) 'Mobbing and well-being: economic and personal development implications', *European Journal of Work and Organisational Psychology*, 5, 203–14.

O'Donnell, T. (1999) 'The sweat of the brow or the breaking of the heart' in A. Morriss and T. O'Donnell (eds) *Feminist Perspectives on Employment Law*, London: Cavendish Publishing.
Robertson vs Forth Bridge Company (1995) IRLR 404.
Sherrif vs Klyne Tugs Ltd (1999) IRLR 481.
UNISON (1997) 'UNISON members' experience of bullying at work', London: UNISON.
Ward vs Scotrail Ltd (1998) SC 255.
Walker vs Northumberland CC (1995) IRLR 35.

PART FOUR

The way forward

CHAPTER EIGHT

A proactive approach

VIVIENNE WALKER

Introduction

In this chapter I will be looking at the way that work on culture and values has contributed to organisational life in a health and social services trust (HSS). Particular emphasis will be given on how the dignity of employees in the workplace is protected.

South and East Belfast HSS Trust is one of 19 trusts in Northern Ireland which provide health and social care to the population. Social care is integrated with health services. As such it is different from NHS trusts in the rest of the United Kingdom where social care is the responsibility of local authorities. The South and East Belfast HSS Trust was formed from a management unit which brought together, for the first time, the areas of south Belfast, east Belfast, Castlereagh and an acute psychiatric hospital. Each area had its distinctive culture and way of working.

The trust currently provides services to a local population of 209,000 people and employs around 3,600 staff based in a range of locations throughout its area. Its responsibility includes the provision and management of community health and social care services to its population through a wide range of facilities including a psychiatric hospital, residential homes for the elderly, hostels for people with learning disabilities, a young people's centre, health centres and clinics.

In this chapter I will describe three pieces of work, which together make up a proactive approach to building a culture, which supports dignity and respect for employees of the South and East Belfast HSS Trust. This has two elements – one relates to respect for the trust as an organisation and the other, respect for the diverse range of people who work in it. They are component parts of a holistic approach which is still developing and changing in the light of experience and feedback from people involved in the processes.

The first piece of culture change work, which I will describe in this chapter, relates to the work on developing and communicating the trust's mission. The main reason for this was a determination to pull the units of management together as a cohesive unit. The other two pieces of work described in the chapter build on the promise made in the mission statement to value staff. The second piece relates to the trust's research study on approaches to making values and expected standards of behaviour clear in a health and personal social services (HPSS) organisation. Finally the chapter deals with the trust's work on dignity in the workplace. While I will go into more detail when describing the trust's work on dignity and diversity I want to make it clear that I do not believe that this could have taken place without the foundation of the work on the mission or the understanding gained through probity work.

I would not want any reader to feel that work in the trust was a tidy process clearly thought out from the beginning. The mission work was the beginning. Implementing supporting systems and processes to the part of it which refers to valuing staff has at times been chaotic – instinctive rather than reasoned – and an uphill struggle. The work we have done is only at the beginning of a long road. That said I would recommend others to start the process.

Organisational background

On the formation of the new unit of management in 1990 (and subsequently the trust in 1994) the new chief executive recognised that there was a need to bring the different groups of employees together and to ensure that they were able to form a cohesive workforce. There was also the practical problem of unifying the different working practices, structures and cultures of the trust's component parts. Secondly, there was the requirement, common to all HPSS trusts, to maintain day-to-day performance and quality standards in the face of major policy and operational changes, which included budgetary reductions and equity shifts. Finally, there was a need to prepare for the new challenges facing trusts and to build positively and creatively on the strengths of the new organisation in order to maximise future performance.

The chief executive initiated a number of programmes to achieve cohesion. One was to develop a mission statement and another to restructure to prepare for the formation of the trust.

Creating a shared mission

Given the need to develop a unified organisation the trust recognised the need for a common focus. To achieve this the senior management team devoted much of its early efforts to developing a clear but challenging mission statement for the trust. This included bringing in senior speakers from a range of external organisations including the private sector to discuss how their organisations achieved cohesion and ensured quality of services.

The resultant mission statement was a statement of purpose supported by an explicit set of values:

> South and East Belfast HSS Trust is committed to provide in partnership with other interested parties the best health and social care for South and East Belfast and where appropriate the wider community through bringing care to people by *valuing* individuals and families, developing *trust* through *involvement*, responding *sensitively* to *needs*; *striving for excellence* by developing *partnerships* with the community, *valuing* its staff, promoting *innovation, research* and *evaluation*, managing for *effectiveness* and *value for money*.

The mission statement was viewed as much more than a paper exercise. It was a powerful declaration of intent that could bring life to day-to-day work. Nearly 3,000 staff took part in workshops linked to the development of the mission statement. The focus of the workshops was to show employees that excellent performance was dependent upon the little things that they did as well as the big strategic decisions made by senior managers. The workshops showed how employees could do their bit to make the mission statement happen. Employees were encouraged to look at their own areas of activity and to identify practical steps that would make mission a reality. Other initiatives, including a significant quality improvement drive, supported and were explicitly linked to the mission work. 'Bringing care to people' became the strap line for all the trust's transport vehicles and as such serves as a constant reminder of the trust's purpose for staff.

Structural change

One of the most significant changes was the removal of the divide between hospital and community. Instead two new service divisions were formed. These were Adult Services and Children's Services. Each of these services took on responsibilities that spanned both hospital and community. In addition two new support divisions were created. One of these was for service development; human resources became a division in its own right.

This major restructuring was the subject of a consultation exercise with about 70 staff who had supervisory or managerial responsibilities. The senior management team led each consultation with a presentation. Each individual involved received a copy of a briefing document which included a letter from the chief executive and a list of principles that were obligatory. Individual comments were taken into account in developing the new structures and follow-up meetings were held for any staff with particular concerns.

The trust's work in this area was published as a case study by the Institute of Personnel and Development (IPD) in 1995.

Organisation values and probity research in the trust

Work on values in organisations is relatively rare. There may be a number of reasons for this. Is it because there is a tendency to shy away from something which is seen as personal and private? Or is it because there may be a fear that if we work with values too much we will arrive at Orwell's *1984*? As Dearlove and Coomber state: 'The very idea that values such as honesty and respect for people make a fundamental difference to the business is not one we are used to' (Dearlove and Coomber, 1999).

It is possible that in our hearts we believe that business, like politics, requires dishonesty or at least shading of the grey to succeed. I believe it is time to revisit that view and that belief was one of my reasons for leading the trust's research into putting probity into practice. This research gave clear indications about how integrating values into organisational life should be approached. It also gave further insight into the effectiveness of the trust's work on its mission statement.

Probity was recognised by the NHS Corporate Governance Steering Group as one of three crucial public sector values. They defined it as:

> an absolute standard of honesty in dealing with the assets of the NHS: honesty and integrity should be the hallmark of all decisions affecting patients, staff and suppliers, and in the use of information acquired in the course of NHS duties.
>
> (Nolan, 1995)

This was reinforced for the HPSS in November 1994 with the issue of guidance, which explicitly referred to probity as a core value.

The probity research undertaken in the trust tested the hypothesis that probity as a concept is not fully understood in the HPSS. The research assessed a number of probity areas including:

- how standards of behaviour were transmitted to employees within the trust;
- what employees understood to be the trust's current position regarding probity;
- the current level of knowledge and the use of probity standards;
- employees' personal needs to behave with probity.

The results confirmed that probity as a concept was not commonly understood. Many respondents confessed that they used the dictionary in order to answer the definition question in the survey. The study also found that there was a low level of knowledge about the existence of the trust's key probity documents and that usage of these documents was even lower. Employees felt drowned by the amount of information passed to them in the daily course of their duties and as a result gave little attention to documents that seemed to have little immediate relevance to their daily work.

Simply put this means that paper-based standard setting runs a high level of risk that the messages about proper standards of service and behaviour are not getting through to the staff who are expected to conform. The issue of circulars giving guidance is common in the HPSS and the NHS. The research indicates that important messages may therefore be missed and hence incorrect assumptions may be being made that changes in practice have occurred.

Table 8.1 Awareness and usage of key documents

Area	Awareness	Usage
The Patient's Charter	75%	48%
The complaints procedure	75%	44%
Confidentiality	56%	36%
Corporate governance	32%	22%

The research showed that the key documents which were best known were those that had received significant coverage in the media. The best-known document at the time of the research was guidance governing the use of management consultants. This had been an issue which received extensive coverage in local and national newspapers as well as on television. Even important recently-released guidance, core to the work of staff, which had been supported by internal HPSS publicity was not as familiar to staff as those areas where scandals had occurred.

The research also showed a significant drop-off between awareness of a key document and actually reading or using it. Some of the research results are given in Table 8.1.

The risks are clear. If staff do not know of or do not read key guidance contained in circulars they will not change to meet new standards of behaviour set out by the centre. They will become out of date and will fail to meet the expectations of users of their services and those who provide funding.

The mission work in South and East Belfast Trust gives an alternative model for communication of change and achieving culture change which works. It is however resource intensive because of the amount of time needed for staff involvement and this needs to be recognised.

The probity research also provided some important reaffirmation of the effectiveness of the trust's approach to developing and implementing its mission statement. Details are given in Figure 8.1: 81 per cent of staff surveyed indicated they believed the trust's mission was relevant to their work, 64 per cent said that the trust's values were evident to them and 77 per cent believed the trust was an ethical organisation. These are critical areas for an organisation working in a health and personal social care environment. They could form an important basis for some benchmarking with other organisations working in the field.

The probity research questionnaire included a section that was based on a research study in 1996, *Managing Ethics*, conducted by the Industrial Society. The trust had been struck by the similarity in the definition of ethical business practice used by the Industrial Society and the issues addressed under probity within the public sector. Many of the behaviours listed in the Industrial Society document as describing an ethical organisation were the same as those used in the HPSS to describe expectations of behaviours and standards of conduct. These included:

Figure 8.1 Mission and values

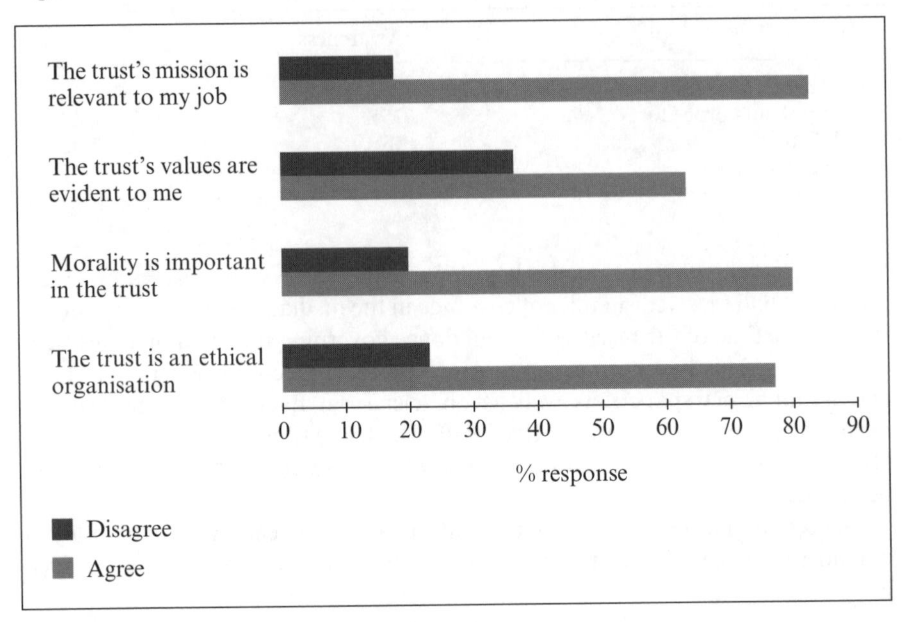

- Open, honest and frequent communication with employees,
- Providing independent redress for employees with grievances,
- Consulting people before taking major decisions that affect them,
- Giving employees clear guidelines on accepting gifts.

Personal discussions with contacts in the Industrial Society showed that their own respondents had also found 'Ethics' difficult to define. Yet, as had been found in the trust, one researcher said during the probity work 'once there was an understanding and exploration of the issues staff readily identify with the subject matter and feel it is a core driver determining how individually and collectively business is done' (Cairns et al., 1998).

The trust found there was a significant conformity of the views in the two surveys. Table 8.2 shows the areas of highest agreement between the two. The comparison between the manager's perceptions of importance given to the same statements by respondents in the trust and the Industrial Society studies was encouraging as it validated those areas the trust considered important.

The research also identified the areas of greatest difference between managers from the trust and managers surveyed by the Industrial Society. Two of these areas are of particular importance to a discussion on dignity at work. Table 8.3 gives some detail. They were:

- observation of the spirit as well as the letter of the law,
- providing independent redress for employees.

Table 8.2 Statements with the highest levels of agreement on importance

Statements	South and East Belfast Trust	Industrial Society
Listening to customer complaints and seeking to resolve them	89%	82%
Respect for people's dignity	79%	72%
Encouraging employees to expand their skills and learning	73%	77%
Giving employees clear guidance on accepting gifts	60%	62%

Table 8.3 Statements with the highest level of difference

Statements	South and East Belfast Trust	Industrial Society
Observation of the spirit as well as the letter of the law	32%	71%
Providing independent redress for employees	48%	73%

The low importance given to these two statements was disturbing to the researchers working in the trust (Cairns et al. 1998). They identified two possible reasons. One could be that the low rating on the observation of the spirit of the law as opposed to the letter of the law was due to the HPSS reforms and the pace of change. Alternatively it may have been due to the ethos that had developed within the HPSS and the NHS, where there is a heavy reliance on procedures and an unwillingness to make decisions in circumstances where rules do not exist.

The low score on independent redress for employees may reflect the traditional approach to employee grievances within the service – that is of sticking rigidly to nationally agreed terms and conditions. These findings may also highlight an increasing alienation of staff, many of whom are highly qualified professionals.

Whatever the reason, these views, if generally held, could be a significant block to the development of a culture in the HPSS and NHS where staff feel valued and respected.

Diversity and equal opportunities

Like many other employers, the trust developed initiatives in the arena of equality of opportunities because of employment legislation. Major developments in employment legislation occurred in Northern Ireland in 1976, some five years after their introduction in Great Britain. Changes continue to be introduced. Employers soon faced intense media interest about decisions and found themselves obliged to pay compensation when cases were found for applicants. Compensation continues to increase. Two recent decisions, directly relevant to how employees are treated in work, have led to compensation levels of £200,000 or above (*People Management*, 2000). Significant changes in the way organisations approach equality of opportunity issues have developed as a result.

There can be a negative consequence to using legislation as a tool to achieve equality of opportunity. People and organisations who are 'punished' tend to feel resentment and many view tribunal findings against them in this light. As a result personal accountability for doing the right thing is avoided. Excuses are sought. This can then develop into an over concentration on getting the procedures right. Equality of opportunity then becomes an issue that relates to protected groups. Groups which are excluded from the legislation find it even more difficult to achieve changes on their behalf. Organisational thinking becomes rigid and inflexible with valuable effort diverted to resisting individual claims of discrimination rather than in an effort directed at preventing complaints occurring. Organisations focus on procedures rather than values and behaviours. It is changed behaviours, not procedures, which reduce the level of complaints.

In Northern Ireland as elsewhere this cycle was reinforced by the activity of the statutory agencies, which often operated as vested interests and vied with each other in an effort to gain prime position. In this environment chasing headlines is an important way of demonstrating success. Winning cases then takes precedence over information, support and advice to employers. The situation was further compounded by the agencies offering different advice for handling similar situations. One notable example occurred in recruitment and selection where one agency supported random selection for short-listing purposes and another opposed it. Differences like this give a hook for people who want to resist change. Effort is directed at complaining about a lack of clarity rather than doing something proactive about the main issue – equality of opportunity.

It will be interesting to see whether the establishment of an Equality Commission by the Northern Ireland Act 1998 will lead to a fundamental difference in approach. The Equality Commission assumes responsibility for all areas of discrimination legislation in Northern Ireland as well as for the new equality schemes required from the public sector. If it is to succeed the new commission must ensure parity in its approaches to all types of discrimination and avoid each specialist area working to its own agenda.

It was experience in industrial tribunals and a general view that 'we can do better', which led the trust to the development of diversity as a concept that could add value. Our view of diversity is based on the idea of a level playing field and is a further step beyond equality. Diversity is based on the premise that people are different and there is value to organisations in recognising difference and channelling it into benefiting the organisation. Dignity, of which I will give more detail later, is in my view a subset of diversity. It relates to the way in which individuals are treated in organisations and bridges both concepts.

The University of Ulster has an enviable track record in its work in this area. It has provided briefing papers on *Western Routes and the 'Civic Leadership in a Divided Society'*. These relate to an initiative on civic leadership with a number of local councils in the province. The theme of the work is equity, diversity and interdependence. The university believes that the time may be right to promote a different model of civic leadership. This demands the creation of an overarching set of values, rights and responsibilities to encourage the full participation of people in Northern Ireland without denying their right to live through and by their different identities. The researchers argue that equality of opportunity has led to a culture of neutrality.

> A culture of neutrality promotes fair employment practice. There are rigorous monitoring procedures and if the numbers do not reflect wider communal break-downs then action is taken – with varying degrees of enthusiasm. Success is gauged in terms of numbers only rather than the state of relationships with differences sacrificed for the sake of harmony . . . Ultimately, neutrality offered both identity groups an opportunity to belong in a mutual relationship based on silence and politeness.
>
> (Eyben et al., 1999)

The University of Ulster is currently taking this concept further in partnership with some local councils and is currently working on the development of diversity standard. I believe work in the trust on dignity has addressed the relationship issues highlighted by the university.

To recap, dignity offers the opportunity of building on the concept of diversity, which as a value base is a strong foundation for moving forward.

Implementing a dignity in the workplace policy and supporting procedures

In 1994 the Trust Board approved the trust's 'Dignity in the workplace' policy. The board recognised that no single tool would be sufficient on its own to bring about culture change. The policy was intended to contribute to the trust's mission statement of valuing staff and was part of an ongoing culture-change programme. A number of organisational support processes were devised to support the introduction of the policy. These included issuing leaflets to all trust employees, identifying and training harassment support officers, adopting formal and informal reporting procedures and developing training for managers. Significant assistance came from managers and trade unions. Further support procedures were developed and implemented in trial form over the next two years. At that time, because of the concentration on equal opportunities, models of best practice in addressing workplace harassment had yet to be established. Human resources staff therefore consulted widely, particularly with managers, external enforcement agencies and trade unions, before deciding upon a preferred approach. An important aspect of the policy was that, while 'dignity' is the core message of the policy, the need for 'balance' is highlighted in its communication – hence the trust aims to achieve a 'harmonious' and 'supportive' working environment.

The main challenge in building the culture of respect was in implementing the policy and ensuring that it 'came to life'. The diversity approach, and its supporting message of dignity for staff, had to be communicated throughout the organisation. It was important that this policy should be regarded as an organisational rather than a human resources initiative and senior executive involvement was explicit from the beginning. A series of managing diversity awareness sessions were organised for managers, with the expectation that they would cascade this message throughout their departments. The managing diversity awareness sessions for senior managers were introduced by the chief executive who communicated the business case for managing diversity, emphasising its strategic importance to the performance of the trust. The trade unions took part in the senior managers' sessions and a mix of human resources and senior executives as well as myself gave information on content. The Equal Opportunities Commission and the Fair Employment Commission warmly welcomed the initiative and undertook the quality assurance of all the 'Dignity in the workplace' documentation. The Labour Relations Agency has also commended the initiative.

During these sessions the senior managers suggested that a training pack should be developed to assist them in delivering the awareness training to their employees and to ensure that the message was communicated consistently and comprehensively across the organisation. Managers played an integral part in developing the policy, procedures and the supporting resource pack. The pack was piloted in conjunction with managers and trade unions and represents a significant training resource that includes guidance on preparing for an awareness session with staff using acetates, speaking notes and case studies for discussion. The trust has also invested in a number of training videos that can be used in addition to the pack.

Harassment support officers were appointed to support implementation. They do so by facilitating and supporting managers in communicating the diversity and dignity message, as well as by supporting individual employees. They received initial training in harassment issues, relevant legislation and listening skills. They were given input for developing their own working practices and standards of service. Harassment support officers were drawn from all grades of staff within human resources and represented men and women from Protestant and Catholic community backgrounds. It is worth noting that an employee seeking support can make his or her own choice of harassment support officer. The option is also open to change Harassment Support Officers. No reason needs to be given and staff have occasionally used this option.

The trust's Staff Care Service was involved in the development of supporting procedures and took part in the training for harassment support officers. (The

Staff Care Service was developed to provide a counselling service for staff on a completely confidential basis. It was then extended to provide services to other organisations that were anxious to provide a similar service for their staff. It has achieved a high profile because of the excellence of services. Counsellors from the trust have been involved in crisis counselling following disasters at Paddington, Omagh and King's Cross.) Two Staff Care counsellors were designated to debrief harassment support officers who may experience stress as a result of particularly difficult cases.

Leaflets on harassment have been issued to all employees in order to increase awareness of the issues and to ensure that the process for making complaints is clearly understood. The leaflets are in simple, clear language. This was achieved by using employees to quality check them for clarity and appropriateness for all levels of employee. One of the roles of the harassment support officers is to ensure that all employees are aware of their right to complain internally and externally. Some general information is given about who to go to and the existence of time limits for making external complaints.

A diversity action plan was developed to identify actions required to consolidate the trust's diversity approach. This action plan is updated regularly and is overseen by a Diversity Steering Group chaired by the chief executive. Examples of initiatives include an Ethnic Minorities Cultural Awareness Day and a presentation to a Chinese women's group, using interpreters and literature translated into Chinese, about employment opportunities in the trust. Workshops to help the understanding of disability and race issues are planned. Diversity training has also been incorporated into the trust's short course programme. Workshops and diversity training focus on raising the levels of understanding. The trust believes that this will lead to establishing common ground and reduce fear of the unknown. Diversity and the unacceptability of harassment have been incorporated into organisational induction. This will ensure that new staff will know about the issue and will reduce some of the risks of diluting the message because of turnover.

Some 30 staff seek assistance from the harassment support service each year. This in itself is an indicator of success. Numbers increase each time there is a communication initiative, such as a leaflet drop, and then decrease which underlines the importance of constantly communicating support systems to staff. It also means different ways of communicating the message need to be developed. The trust is currently using leaflet drops, messages on pay slips, posters for notice boards and various meetings for this purpose.

The trust's staff satisfaction survey carried out in 1998 gives some measure of the success of the dignity in the workplace policy and procedures. While direct comparisons with other surveys are difficult because of differences in definition and time they can give some broad indications.

In the trust's survey 71 per cent of staff felt they had been treated fairly. These findings were shared with personnel staff at Bombardier, a large local employer in aircraft and missile manufacturing, with whom the trust was benchmarking personnel processes. Indications were that the trust's results were significantly better than those obtained in their survey.

As the trust is based in Northern Ireland, there was a danger that an emphasis would be placed on the issue of sectarian harassment. This did not prove to be the case. The staff satisfaction survey asked specifically if staff felt they had been harassed at any stage during employment with the trust. One hundred and eighty-nine respondents – just over 18 per cent – indicated they had been harassed. One hundred and eighteen of these – just over 14 per cent – indicated this was due to bullying. Managers and colleagues were reported to be perpetrators and were reported at almost equal levels. Staff also reported a significant degree of bullying on the part of users of services. I find these levels worrying. However I take some comfort from other surveys. A survey by UNISON (1995) found levels of around 20 per cent and a Royal College of Nursing (RCN) survey (1997) found levels of around 33 per cent. Research into bullying in an NHS community trust by the *British Medical Journal* indicated that 38 per cent of staff responding reported being subjected to bullying behaviours (Quine, 1999). It also reported that support at work might be able to protect people from some of the damaging effects of bullying. It is early days but it may be that the work in the trust on support systems may account for the lower levels. The survey also asked staff how they rated support systems in the trust and 61 per cent of staff who felt they had been harassed stated support systems were effective.

As a result, the trust has paid attention to the issue of workplace bullying. An awareness session has taken place for managers to consider specifically the characteristics and consequences of bullying. This was organised in partnership with UNISON. A leaflet has been produced for staff defining what bullying is and setting out ways of stopping the bullying.

Anecdotal evidence indicates that there is a heightened awareness amongst staff of the need to be sensitive towards others, both colleagues and clients. One quote from an evaluation sheet reads: 'It really made me realise how important it is to be sensitive to other people's feelings.' One manager says: 'I use the pack with my staff. It opens their eyes to the value and contribution of people from different experiences and backgrounds. I believe it helps increase the quality of service we provide to our client group.'

There is also evidence that staff appreciate the availability of a support mechanism. They say:

> If I had not had the support I would no longer be with the trust. Since the issues were resolved I have been very happy at my work. I feel I have been listened to and what I said was important. I know the support is there for me.

> You can have no idea how much independent support meant to me. I could not have raised how I felt with my manager. As a senior manager I feared it would be taken as an indication I was not able to deliver professionally and that I was a silly emotional woman.

In addition a number of external organisations have contacted the trust for assistance in developing their own harassment policies and have used the 'Diversity presentation pack' in carrying out managing diversity awareness training with their staff.

A formal review of the policy and procedures has taken place and amendments have been made to streamline the systems. Changes have been made in light of the experience and bullying is now specifically included. Consideration is being given to develop an internal conciliation and mediation capability. Another development is the extension of harassment support to include a support service for staff experiencing domestic abuse. This work is being done in partnership with the biggest union in the trust, UNISON. There are six dedicated support officers in the trust. Three of these are harassment support officers and three are trade union representatives. The service aims to provide first-line support and practical assistance in order to help staff deal, in the way they choose, with the situation in which they find themselves. Part of this will be to provide a link to professional services if the individual involved requests it.

Once the revised policy and procedures are communicated to staff, a long-term strategy of communication will be devised.

Learning from initiatives

The message from the mission work is unambiguous. Face to face communication – particularly when it involves senior people in the organisation – does work. The reason for this success was identified as the belief that if a senior manager is prepared to give time to communicate a message personally then that message must be very important. Feedback indicates that systems of intensive communication do result in ownership of messages and better understanding. Finally, the work in the trust has demonstrated it is possible to make mission statements live in organisations.

The message the trust received from the probity research is also unambiguous. Paper-based approaches, which are the traditional means of communication in the public sector, do not work. They do not work because of their volume, their bureaucratic 'all things to all men' language and because most busy people – particularly those working at the front line of an organisation – can find other things to do rather than spending time reading about someone else's expectations of their behaviour.

The findings from the probity study and the mission work are of direct relevance to building a culture of dignity and respect for others in an organisation. Two of the themes which emerged are particularly important. The first is that the manner in which messages are put across is crucial to people's understanding of the whole. Written messages are not necessarily the most effective. The second is that there are subtle and sometimes significant differences in context. If organisations do not take context into account, messages about values will not appear real to those who receive them. Finally the process itself – and the discussion generated by it – may be a vital catalyst for improvement. Any work on building a culture of respect requires a personal approach which involves the most senior managers.

Work on general values is an essential precondition to any work on staff rights in organisations because this establishes the ground on which a culture of respect can be built. Work on a culture of respect is sensitive. It exposes our most private and deep-rooted prejudices. It therefore needs a strong foundation or it will fail. It also needs time and a determination to keep going.

One of the problems in undertaking the 'Dignity in the workplace' project was equipping managers with the confidence to undertake the cascading of the diversity awareness message throughout the organisation. Many managers perceived that this would require human resources expertise. This problem was overcome through the development of the presentation pack and the provision of assistance from harassment support officers. Human resource departments need to be aware that the reason that initiatives come to rest with them may sometimes be because managers are uncertain about how to go forward and not that they resist the message.

The rise in and fall-off in cases where staff come forward for harassment support following communication initiatives illustrates the importance of sending out the message that harassment and bullying is unacceptable. Organisations cannot relax once an intensive communication initiative is complete. Reinforcement is a crucial part of implementation.

Benefits to the trust from the three pieces of work

Staff identification with the mission statement is one of the major benefits to the trust. This ensures staff have an awareness of the basic principles by which the trust wants services delivered. In addition senior executives in the trust believe the work on the mission statement, and the cohesion which resulted, helped the trust reduce the organisation turbulence caused by the major changes it experienced. These included significant equity shifts of resources from the trust to others within the Eastern Board area and ward and facility closures occurring as part of the shift of care into the community.

Feedback from staff involved in the probity research indicated they were grateful to the trust for involving them in an exercise which would help give clarity to their day-to-day work and the dilemmas they faced when doing it. In a practical way the trust was demonstrating its determination to value and respect staff. In addition the trust now understands the dilemmas staff face more clearly. This helps when changing work patterns or re-engineering.

Understanding differences is helping the trust value its staff. This helps improve services because the trust understands the needs of a wider range of people. The trust has a profile in Northern Ireland of being proactive about diversity and equality of opportunity. This has helped in attracting some managerial recruits.

A level of expertise has been developed in dealing with some of the more difficult employee relation matters in the trust. The trust's expertise in dealing with issues of harassment has led to a significant reduction in the number of cases going forward to industrial tribunals where this is the basis for a complaint. Success in dealing with internal problems, which could lead to applications to go to tribunal, is considerably improved.

Staff willingness to be involved is clear from all three initiatives and is an important benefit to the trust because it helps in achieving organisation change.

Conclusion

Some word pictures catch our attention in ways others do not. I am much taken by the concept that life is like a flame. It is a concept used in the publicity of the Marie Curie Cancer Fund. Elton John used it, to great effect, when he revamped *Candle in the Wind* as a tribute to Diana, Princess of Wales, following her death. To me flames grow if there is enough 'food'. They die if there is not. They grow if there is room to expand. They die if there is not. I believe this reflects life in organisations. Values and the resultant culture that binds them are the food. Managers are charged with providing the room to grow and the support to ensure effort is directed appropriately.

To some readers this may appear trite, however experience and learning in the trust indicates that values are the timeless guiding principles which influence everything the organisation does. They are the invisible force which can make or break strategic and operational decision-making. They are not necessarily uniform across an organisation. They can help or hinder managers in their efforts to achieve alignment between organisational and individual objectives. They can, if developed, supported and used appropriately, reduce the turbulence of change and increase the organisation's ability to meet a crisis.

I believe work on values is the necessary precursor to developing a culture of respect for and between the people who work in it. This view is supported by the benefits which the trust has seen from the work it has done. The trust is not alone in its belief.

Professor John Kay supports this view in a lecture that he gave to members of the RSA. His hypothesis was that profit is not and cannot be the sole motivator of business. He states 'successful business is not in reality selfish, narrow and instrumental' (Kay, 1999).

He describes how good business is multi-dimensional and complex and gives examples of how values and concern for other members of a team contribute to successful long-term business operations. He goes on to say:

> these statements do not truly represent the values of successful businesses. These are organisations which serve the needs of their customers, provide a rewarding environment for those who work in them, satisfy the requirements of those who finance them, and support the development of the communities in which they operate.
>
> (Kay, 1999)

Further support for the view is found in a four-year study of between nine and ten firms in each of 20 industries conducted by Kotter. The researchers found that firms with a strong culture, based on a foundation of shared values, outperformed the other firms in the study by a huge margin. 'At a deeper level, corporate culture is about the implicit shared values among a group of people – about what is important, what is good and what is right' (Kotter and Heskett, 1992).

Further support from a UK perspective comes from the Sheffield 'Effectiveness' programme. Researchers in that study attributed 29 per cent of the variation in productivity over a period of three or four years to the human relations of culture. A concern for employee welfare was the single most important predictor (*People Management*, 1999).

I sometimes think that agreeing policies and procedures to support a culture of dignity and respect is the tip of the iceberg. In itself the experience in the trust is that it is an immense piece of work. However when the effort needed to implement is factored into the equation this pales into insignificance. Implementation is the bit of the iceberg which is hidden under the water. Perhaps that is why so many well-meaning policies lie gathering dust on shelves in personnel offices. The commitment of the chief executive and senior management is what makes the difference.

References

Alderman (1997) 'Bullying in the workplace', *Nursing Standard*, 11, 35, 22–6.

Cairns, J., Hewitt, I. and Walker, V. (1998) 'Probity into practice', research study, Belfast: South and East Belfast HSS Trust.

Dearlove, D. and Coomber, S. J. (1999) *Heart and Soul: A Study of The Impact of Corporate and Individual Values on Business*, KPMG.

Eyben, K., Morrow, D. and Wilson, D. (1999) *Western Routes and the 'Civic Leadership in a Divided Society'*, Belfast: University of Ulster.

Health and Social Services Executive (1994) *Code of Conduct, Code of Accountability*, HSSE.

Industrial Society (1996) *Managing Ethics*, Best Practice Guides, Industrial Society.

Kay, J. (1999) 'Ethics and the role of business in society', *Royal Society of Arts Journal*, 3, 1999.

Kotter, J. P. and Heskett, J. L. (1992) *Corporate Culture and Performance*, Free Press.

Nolan, The Rt Hon. Lord (1995) *Standards in Public Life: First Report of the Committee on Standards in Public Life*, HMSO.

People Management (1999) 'Realising our assets', *People Management*, 14 October 1999.

People Management (2000) 'Stress damages top £200,000', *People Management*, 20 January 2000.

Quine, L. (1999) 'Workplace bullying in NHS community trust: staff questionnaire survey', *British Medical Journal*, vol. 318, 23 January 1999.

UNISON (1995) *Violence at Work*, health service staff study, London: UNISON.
Walker, V. and Hewitt, I. (1998) *Staff Satisfaction*, internal document, Belfast: South and East Belfast HSS Trust.
Walker, V., O'Hara, R., Mulholland, R. and Higgins, M. (1997) *Diversity Presentation Pack*, internal document, Belfast: South and East Belfast HSS Trust.
Walters, M. (1995) 'Organisation culture in public sector organisations', *Issues in People Management*, no. 10, IPD.

Legislation:

Equal Pay (NI) Act 1970 (amended 1984).
Sex Discrimination (NI) Order 1976 (amended 1988).
Northern Ireland Act 1988.
Disability Discrimination Act 1995.
Race Relations (NI) Order 1997.
Fair Employment and Treatment (NI) Order 1998.

CHAPTER NINE

A total quality approach to building a culture of respect

NOREEN TEHRANI

Introduction

Most organisations aim to be successful and to build a reputation for excellence in their chosen sphere of activity. This desire for excellence is found in commercial organisations as well as in public organisations and charities. It has been recognised that a total quality approach in organisations has led to improvements in innovation, the delivery of services and value for money resulting in significant benefits for organisations (Deming, 1982). However there has been less emphasis on a total quality approach being used to create excellence in organisational environments or cultures. This total quality approach recognises, values and respects the unique contributions of each of its employees and stakeholders. Achieving world class performance in terms of respectful attitudes and behaviours does not happen by chance but requires an active process involving the participation of everyone in the organisation from the chief executive to the most junior worker (Kano, 1993). The process of building a culture of respect although often difficult and frustrating is always worthwhile and rewarding.

The aim of this chapter is to show how excellence within the organisational culture and behaviours can be achieved in much the same way as other quality improvement targets. The benefit of adopting this approach to bring about organisational change is that the total quality tools and techniques are already developed and have demonstrated their ability to achieve improvements in individual and organisational performance. All that is required by the organisation is to redefine and adapt these tools and techniques and to focus their efforts on achieving excellence in attitudes and behaviours between and towards individual employees. An organisational case study illustrates the key points discussed in the chapter.

Establishing a continual improvement approach

Although there are many approaches to the achievement of world class standards in the creation of an organisational culture of respect they rely on a number of fundamental elements. The following five elements are based on the elements most commonly found in quality initiatives:

- creating an atmosphere in which the organisation and its leaders have a clear vision of and a sense of what a culture of respect would be like in practice;
- establishing and integrating a continuous improvement approach that is built upon the shared belief that change does not happen by chance but is made to happen;
- developing monitoring tools that measure qualitative and quantitative improvements in the culture of the organisation;
- identifying the necessary tools and approaches required for maintaining the momentum of continual improvement;
- achieving recognition for success through benchmarking and assessment by external bodies such as 'Investors in People' and 'The European Foundation of Quality Management'.

It is unrealistic to expect that people in organisations, or the organisations themselves, will behave in ways that respect the dignity of the individual. Behaviour within organisations is influenced by a number of forces. These include the power of the organisational culture, the pressure of the working atmosphere and the strength of the individual's values and beliefs. Changes in the direction and momentum of any of these forces take time and effort if the change is not to be illusory and temporary (Clarkson, 1995). The importance of the role of senior management in leading a total quality culture change process cannot be over emphasised (Joyce, 1995; Thiagarajan and Zairi, 1997a). It has been found that where senior managers have given lip service to the idea of creating a culture of respect and not taken the necessary steps to change their own inappropriate behaviours, the initiative has failed. Employees quickly become cynical when faced with mission statements that exhort them to behave in a way that bears little relationship to the actual behaviours of the managers they observe in their daily life. Cultural changes require the constant endorsement of all levels of managers particularly the most senior managers. Where the new behaviours are not reinforced by the constant enactment of new behaviours throughout the organisation the cultural change has rarely been successful (Bertram, 1991). Often the problem is caused by the responsibility for cultural change being delegated to middle managers, quality departments or external consultants instead of being the responsibility of all employees.

A number of leadership competencies have been identified as important in building a climate of change (Joyce, 1995). These leadership competencies include relating to people, personal integrity, visibility, commitment to excellence and willingness to challenge the status quo. Senior managers and other leaders can check their competencies in a self-assessment questionnaire (see Figure 9.1).

Even the most dedicated leader committed to achieving a quality management approach to cultural change must recognise the time and effort that will be required to maintain the momentum of the approach. Changing the organisation's culture cannot be achieved overnight but will require the leader's continued interest and involvement over a number of years. Leaders need to be seen championing the culture of respect by continually discussing the subject with employees, by providing formal and informal recognition for the achievements of employees, by receiving training and providing training for others in such areas as interpersonal communication, equal opportunities and diversity. Senior managers will also need to develop and communicate the importance of engaging in a programme of continuous improvement, team working, management by fact, mutual respect and dignity and the value of individual employees (Easton, 1993). What is required is a new breed of leaders with personal integrity and sense of identity.

While the involvement of senior management is vital to the success of a total quality initiative it is not sufficient on its own (Haksever, 1996). There is a need to involve the entire workforce and to gain their commitment. It is only with universal commitment that the momentum of change can be maintained (Crosby, 1989). This commitment is maximised when the workers themselves are actively involved in identifying the areas for improvement and in setting the goals and standards of behaviour. The involvement and empowerment of front-line workers can only be achieved with the active support of middle managers. It is the middle managers who have the greatest opportunity to demonstrate that a culture that respects the dignity of the individual employee is possible. Where middle managers believe that they can only achieve business objectives through an authoritarian or coercive approach they are likely to react with suspicion and resistance to a process of change that would involve a significant change in their own behaviour (Manz and Sims, 1993). It is important therefore that senior management provide the encouragement and training to enable middle managers to become leaders within this new culture. Although management training will be aimed at increasing knowledge of the quality processes and providing the necessary interpersonal skills, it is the people skills that are the most challenging to managers who find the transition from management by coercion to one of increased teamworking and autonomy difficult. For some existing managers these skills may require a trainer or counsellor to undertake the task of identifying and responding to underlying beliefs and fears which may prevent the manager from adopting a more flexible style of leadership.

The quality improvement cycle

The process of quality improvement requires control and this needs to be accommodated and assimilated within the existing systems and processes. This cannot be undertaken quickly as it will require all the existing practices and

Figure 9.1 Commitment to building a culture of respect

In order to identify your commitment to building a culture of respect, please complete the following questionnaire.

Characteristic or behaviour			
People focus	*Sometimes*	*Most of the time*	*Always*
Do you give people personal responsibility?			
Do you actively seek out the views of others?			
Are you committed to team development?			
Do you instil confidence in others?			
Do you encourage open feedback and debate?			
Personal integrity	*Sometimes*	*Most of the time*	*Always*
Do you do what you say you will do?			
Do you show respect to everyone?			
Can you say sorry when you have made a mistake?			
Are you open and honest about your mistakes and learn from them?			
Are you fair in all your dealings with others?			
Visibility	*Sometimes*	*Most of the time*	*Always*
Do you actively promote an 'open door' approach?			
Do you champion a culture of respect and dignity?			
Are you 'available' to listen to the views of others?			
Are you prepared to talk to customers and clients about the need for respect and dignity at work?			
Have you put building a culture of respect on the organisation's main agenda?			

Figure 9.1 (continued)

Promoting excellence	*Sometimes*	*Most of the time*	*Always*
Do you establish individual and team goals?			
Do you give personal recognition to others?			
Do you use feedback and coaching constructively?			
Do you schedule regular time to improving interpersonal relationships?			
Are you constantly looking for opportunities for improvement?			
Challenging the status quo	*Sometimes*	*Most of the time*	*Always*
Do you openly challenge unacceptable behaviour?			
Do you seek out prejudiced attitudes?			
Do you critically examine policies and procedures to make sure that they are fair to everyone?			
Are you committed to equal opportunities for all?			
Do you actively support diversity in the workplace?			
How did you do?	*Sometimes*	*Most of the time*	*Always*
Total scores			
For each score multiply by the following weighting factor	0	2	5
Add these totals			
Total score			
Target			*125*

procedures to be critically examined to ensure that they meet the requirement to respect the dignity of all employees. Although personnel policies and procedures such as those relating to reward, recognition, grievance resolution, recruitment, training and development are important the process does not end there. Organisational policies and procedures should also be critically examined. These policies and procedures may include the allocation of duties, shift working patterns and management styles which are equally liable to disadvantage or harm vulnerable employees. Deming (1982) developed a quality improvement cycle that has been adopted by many organisations as a systematic process in the development of lasting improvements. This cycle has four simple stages that form a framework within which a disciplined approach to solving quality problems and implementing lasting solutions can be identified. The four stages are planning, doing, checking and acting.

The quality improvement cycle requires a structure in which to operate. This normally involves a team of people who would take on the responsibility for managing a specific project using the quality improvement cycle. In the planning phase of the cycle, the team needs to establish the facts, look for any underlying reasons for why the problem exists, identify potential solutions and agree the criteria for success. A comprehensive plan to implement the solutions is then developed, together with a timetable, stages and milestones. The 'do' phase of the cycle involves the implementation of the plan including the on-going monitoring of the outputs of the plan. In the 'check' phase the results are evaluated against the plan's objectives and the benefits measured. Where the plan is not successful, or fails to reach the anticipated level of achievement, the reasons are determined. In the final stage of the cycle corrective actions are taken to address any under-achievement or failures. Where there are gains achieved the process is made permanent and the learning gained is communicated. It is this learning which is at the heart of the organisation's ability to adapt to a rapidly changing environment (Prokesch, 1997).

In a study of groups using the control cycle (Collard, 1993) it was found that where these groups were unsuccessful, this was caused by a number of common factors. These factors included a lack of time to undertake the planning, implementation and review, a lack of organisational commitment to implement the group's recommendations and the selection of over-ambitious or over-complex projects. Groups can also have difficulties if they do not consult and communicate effectively with the entire workforce on the purpose and benefits of the projects that they are undertaking. Successful projects tend to include some 'quick-win' elements that build team confidence to tackle more complex issues.

Case study: project introduction

A large distribution company had become aware of problems that were occurring in one of its depots. This problem had become more acute when women had been introduced into the workforce. The women complained about the leering and sexual comments being made by the men. It then became clear that this was only the tip of the iceberg and that there was a culture in the depot that encouraged the use of bullying tactics by both supervisors and other powerful members of the workforce. Local managers were aware of the problems but did not feel able to deal with them without becoming the victims themselves. The senior managers were convinced that this culture was not only destructive to the individuals who were being bullied or harassed but also that it prevented improvements to the productivity of the depot.

It was decided that there should be a number of focus groups to look at what was happening and to identify the ways of dealing with the problems. The focus groups were made up of a mixture of senior managers, middle managers, employees and the union. The groups focused on how the members of the group would like to be treated by their fellow workers and also provided an opportunity for each group to describe the problems that they faced as part of their role. Each group developed a code of acceptable behaviours (Tehrani, 1996) and then went on to describe how these behaviours related to the work within the depot. The groups then identified how they would deal with individuals who breached the code of acceptable behaviours.

Monitoring benefits and improvements

There is a well-known tendency for people to only do those things that are rewarding. In organisations rewards are based on features of work that can be measured or evaluated. This is quite difficult to achieve when the feature is an interpersonal behaviour or where the change is qualitative as is the case with changes in attitude or approach. When applied to production and the delivery of services quality, management uses hard data such as a cost benefit analysis where it has been shown that significant savings are possible (Joyce, 1995). The cost of introducing a total quality approach to cultural development that includes introduction and monitoring standards is relatively easy to calculate, however there are more difficulties in calculating the costs of failure. The benefit of an excellence in the quality of the interpersonal relationships can be difficult to evaluate quantitatively. A number of indicators that can provide some hard evidence on the effect of introducing a culture of respect are available. An example is to look at the number of cases taken to employment tribunals, changes in the use of the formal and informal grievance procedures and feedback from employee satisfaction surveys. One way to assess changes in culture is to identify key indicators of the desired change and to ask employees about the progress towards the identified goal. A sample of an employee questionnaire is shown in Figure 9.2.

Although it is difficult to quantify the indicators in Figure 9.2 precisely, the qualitative information on the transformation of management roles and

Figure 9.2 Building a culture of respect: how have we progressed?

This questionnaire has been designed to identify the extent to which you believe that the organisation has changed following the introduction of the culture of respect. Please think about these indicators of change and provide your comments in the space provided.

Key: Change for the better 1 No change 2 Change for the worse 3

Indicator	Change	Comments
Does the senior management team take ownership for some of the problems and bring about real change?		
Does everyone in the organisation have a clear understanding what it means to have a culture of respect?		
Is the culture a subject that is regularly discussed in team meetings?		
Do you have clear and measurable targets for achieving a culture of respect?		
Are individuals able to challenge unacceptable behaviours in others without fear of retaliation?		
Do you feel that there is a basis for openness and trust in your team?		
Is there anything else that would help to build a culture of respect?		

behaviours, together with workers' attitudes and motivations, provide feedback to show that things are changing for the better (Ghoshal and Bartlett, 1997).

Case study: monitoring the benefits

The data gathered in the focus groups was collated and an agreed code of acceptable behaviour was produced and communicated to the workforce. The focus groups had also developed a process for dealing with behaviour that was unacceptable to the groups. This process recognised that it was the responsibility of everyone to identify breaches of acceptable behaviour to the individual concerned. This 'owned' approach to maintaining a culture that had been defined and agreed by the groups was found to be effective in reducing the number of complaints of bullying and harassment.

Tools for continual improvement

Initiation

A number of quality management tools are available when considering the initiation of a project to deliver a more respectful organisational culture. An overview of each of the tools is provided as a starting point.

Focus groups

A key to the success of a total quality approach is the use of focus groups or quality circles to identify ways of achieving a systematic approach to shared problem-solving. While individual groups vary in their style of operation there are some common principles that must be applied if the group is to achieve its goals.

- depersonalising conflicts through separating personalities from problems by adopting a systematic approach;
- providing a logical framework which stimulates the emergence of relevant facts as the only determinant of solutions;
- clarifying and integrating the needs and objectives of the organisation with those of the people working in the organisation.

Focus group decision-making is based on an active problem-solving process (Collard, 1993). This process ensures that discussions are carried out in an organised and controlled way in order to achieve specific results. The group needs to be managed by a facilitator trained in the skills of group dynamics, brainstorming and a wide range of interpersonal skills such as questioning, building, supporting, clarifying and challenging. A checklist (see Figure 9.3) gives a basis for running a focus group. When a focus group is correctly constructed and motivated it is possible to achieve concrete results and to have the means to demonstrate the process that led to a particular decision.

Brainstorming

Brainstorming is a technique that is used by focus groups to encourage creative thinking and to enable a large number of ideas to be generated in a short space of time. Commonly it is used at the beginning of a focus group session when there is need for a divergent approach. There are a number of rules to brainstorming, one of the most important is that all ideas are equally valid and no one should be treated in a way that undermines or diminishes their contribution. Brainstorming when used to elicit a wide range of ideas from the group will require a facilitator to control the session and to write down all the ideas.

Figure 9.3 Focus group checklist

Questions	Response
Have you agreed the nature of the problem? What is it?	
Have you established the key facts? What are they?	
Have you restated the problem in view of the key facts? Is there a difference?	
Have you identified the potential obstacles to solving the problems? What are they?	
Have you identified ways around the obstacles? What are they?	
Have you established the criteria against which the solutions can be measured? Is the measurement criterion practical?	
Have you agreed an action plan? Does it include a communication process?	

Brainstormers will be encouraged to take a 'blue-sky' view of the issue being brainstormed as the most unusual or apparently outrageous ideas often are the ones that eventually lead to a solution. At the end of the session the ideas can be evaluated by a process of rational analysis.

Employee surveys

Self-administered employee questionnaires and surveys are a quick way of obtaining information from a large number of employees. However there has been a tendency for this tool to be misused. Often surveys are badly designed and as a result provide little useful information. When designing a questionnaire it is important to be clear about the information required and that the information is reliable and valid. The design of surveys requires special skills (Oppenheim, 1966) and the author needs to be aware of the pitfalls that are encountered in poorly constructed surveys. Some of the most common errors involve the use of leading questions that suggest the desired answer, an example of this is 'most people believe that bullying is unacceptable. What is your view?' Clearly people tend to follow the direction of the leading statement. Designers of employee surveys also need to make sure that their surveys are attractive to the user. It takes time to design an attractive layout and to ensure that the questions are easy to follow. Many of these difficulties can be eliminated with the carrying out of a pilot survey that identifies areas of ambiguity or confusion. The instructions on how the survey is to be completed and what will happen to the information gathered should be clearly described on the survey. This is particularly important when the information being sought is of a sensitive nature and the confidentiality of personal information is maintained. Employee surveys are very useful in the initiation phase of a project as they can provide a snapshot view of employee opinions and attitudes. Surveys can also be used to provide feedback on initiatives and, if the structure of the survey is maintained over a number of administrations, the results can be used to monitor changing attitudes and opinions.

Graffiti boards

Graffiti boards provide a very quick and easy way of generating ideas and eliciting employee views. This technique involves the placing of a flip chart or whiteboard in a prominent position such as next to the staff restaurant or coffee machine. The question or issue for discussion should be clearly identified and employees should be encouraged to write their comments on the board. Comments may be anonymous or signed. Each day or week the comments are recorded and considered by a graffiti board project group. The group has the task of collating, considering and responding to the comments. The comments and the responses are then circulated. This technique has been shown to be helpful in identifying underlying issues and problems, but its use and effectiveness is diminished if the process of responding to the comments is delayed or inappropriate. Some graffiti board issues could be: 'What are the most important ways in which the organisation can demonstrate that it respects the dignity of individuals?' 'Which policies or procedures could be changed to improve the ways that people communicate with each other?' 'If you were talking to the MD, what would you ask him to do to make the climate in the organisation more respectful?'

Action

Managing by process

Organisationally, process improvement is a way of facilitating changes in organisations. A process map graphically represents an organisational process or activity. Process maps are useful in establishing the way that organisations handle people management, policies, strategic development and resources. Managing by process frequently cuts across traditional functional boundaries to achieve the best outcome or result. Where organisations maintain a strict functional approach to problem-solving there tends to be less understanding of how individual actions and behaviours affect the whole organisation (Thiagarajan and Zairi, 1997b). When introducing a culture of respect each organisational process will need to be defined and then checked to ensure that the process itself is consistent. This fundamental principle is encapsulated in the European Foundation for Quality Management model (EFQM). This model proposes that results are only achieved by the involvement of all employees in the improvement of processes. A useful way to look at process maps is to put yourself in the role of someone using the process and then to ask yourself how you would feel in that situation and what other information or support you might require.

Case study: examining the processes

A female employee was sexually harassed by a senior manager and raised the problem with her line manager. During the meeting with her manager it became clear that the harassment was being taken very seriously and immediately after the meeting the line manager informed personnel and started the procedures for dealing with the harassment case. The employee was never asked how she wanted to proceed and soon found herself being interviewed under the formal discipline procedures. At no time was the woman asked how she would like to deal with the situation and as a result she felt not only that the senior manager had harassed her but also that she was being bullied by the organisation into taking out a formal grievance action. A small change to the harassment procedure, which involved explaining her options and allowing her the right to choose how she would deal with the problem, would have saved her from this distress.

Conciliation and mediation

The use of conciliation and mediation to deal with bullying in the workplace is relatively new in the area of bullying. Although often used as interchangeable terms there are some significant differences between mediation and conciliation (Reynolds, 1997). Mediation involves the use of reasonable discussion between the people involved. The mediation process is led by an impartial person who facilitates the meeting of those involved with the aim of coming to an agreement. In mediation it is the people involved in the situation who decide upon the terms of the agreement. The mediator's role is to facilitate this joint agreement and not to offer any advice or solutions.

Conciliation is rather similar to mediation in many respects, however in conciliation, the conciliator uses discussion, persuasion and communication skills to help in the negotiation of a solution (Debell, 1997). Unlike a mediator, a conciliator will frequently offer information, advice and possible solutions to the people involved in the process. The advantage of using conciliation and mediation in resolving bullying cases in the workplace is that, unlike other methods, the process allows people to discuss their dispute with a trained person and then to focus on common goals. In most cases people do not understand the full extent of the effects of their behaviours on others, the mediation and conciliation processes allow this understanding to be achieved in an atmosphere which is fair and balanced.

Mediation and conciliation are voluntary processes and cannot be imposed. These processes are involved in helping people to come to a mutually acceptable agreement. This is different to disciplinary or grievance processes which are based on the adversarial process which leads to winners and losers.

Monitoring

360-degree feedback

An effective way of monitoring changing behaviours is through the use of 360-degree feedback. This requires a particular behaviour or attribute to be selected and then the views of peers, managers and teams are sought on the extent to which an individual meets the standards set. The 'Commitment to building a culture of respect' questionnaire outlined in Figure 9.1 could be used to monitor the achievements in a 360-degree feedback exercise.

Cost benefit analysis

Although there have been a number of attempts to quantify the costs and benefits of investments in employee health and well-being, in practice these have proved rather difficult to implement within organisations (Mossink, 1997). A model developed by Oxenburgh (1997) looks at the costs of implementation of a programme and then the changes that take place following the implementation. The key cost/benefit factors that are examined in the Oxenburgh model are the number of productive hours worked, the wages or salary costs, employee turnover and training costs, productivity and quality shortfall and the cost of the programme. When considering a programme that is designed to create a culture of respect, the cost of failure needs to be included. This would involve looking at the time taken to deal with discipline and grievance cases, the legal costs related to employment tribunals and other legal actions, the management time taken in defending the organisation from litigation, and the cost of bad publicity relating to industrial action.

Benchmarking and best practice

Benchmarking is a process that compares the performance of one organisation against the performance of another. The aim of the benchmarking organisation is to look for a company which is doing something well and to learn how it is doing it in order to emulate the process (Manganelli and Klein, 1994). Benchmarking

Table 9.1 Benchmarking guidelines

- Prepare well in order to make the most of the benchmarking partner's time and to prevent legal problems.
- Use designated contacts and show respect to the benchmarking partner's organisational culture.
- Provide the same type and level of information in return.
- Maintain the confidentiality of the benchmarking partner.
- Only use the benchmarking information for the agreed purpose.
- Ensure that the benchmarking does not conflict with the law (e.g. competition or compliance law).
- Provide the benchmarking report as agreed with partner.
- Respect your partner's views on how they expect you to use the data.

can be problematical if not undertaken appropriately. Table 9.1 provides guidelines for organisations contemplating undertaking a benchmarking exercise.

There are some simple rules to benchmarking; these include building a climate of trust, because without trust benchmarking partners will not share information. Joyce (1995) describes a systematic benchmarking model that involves four steps: planning, collecting data, analysing and adapting and improving. This model begins within the organisation and only involves benchmarking partners when the area for benchmarking has been researched internally and the areas for benchmarking identified. An alternative approach is to work with a number of interested benchmarking partners to define the area for research and then to carry out the research concurrently.

Self-assessment

A highly effective quality tool is the use of self-assessment. In the self-assessment process groups of employees will examine the way that the organisation is performing against agreed performance standards. This assessment will not only look at the organisational approach but also at the way that the approach is deployed throughout the organisation. The objective of the assessment is to identify the strengths and areas for improvement. The results of the self-assessment can then be used to identify the priority areas for action, based on the negative impact of issues on employees and the organisation. Figure 9.4 provides a framework for a self-assessment discussion on the achievement of a culture of respect.

Recognition of achievement

Internal recognition

The importance of recognition cannot be overestimated in the achievement of excellence. Organisations that use reward as an incentive to advance quality

Figure 9.4 Self-assessment on the achievement of a culture of respect

Criteria	Score*	Comments
How leaders visibly demonstrate their commitment to a culture of respect Develop clear values and expectations for the organisation Act as role models Give and receive training Make themselves accessible to listen and respond to ideas and concerns Are actively involved in improvement activities *How leaders support the culture of respect by providing resources* Define priorities Fund improvement activities Use appraisal system to support improvement and involvement *How the leaders' behaviours promote a culture of respect* Recognise the efforts of individuals and teams in creating the culture of respect Promote a culture of respect with customers, suppliers and on other appropriate platforms *How the organisation develops policies and strategies based on appropriate information* Gathering information from employees Consulting with external organisations Benchmarking Considering social, health and legal issues *How policy and strategy, related to the organisational culture, are developed* Is responsive to the need to show respect to all employees Involves employees in developing policy and strategy *How the culture of respect policy is communicated and implemented* Uses the policy to plan activities and set objectives throughout the organisation Evaluates, improves and prioritises plans Evaluates employee awareness of the policy and strategy		

Figure 9.4 (continued)

Criteria	Score*	Comments
How policy is updated and improved Evaluates the relevance and effectiveness of the policy and strategy Reviews updates and improves policy and strategy *How the organisation uses the culture of respect to release the full potential of all its people* Ensures fairness in terms and conditions of employment Shows respect in the handling of issues relating to working hours, remuneration, redeployment, redundancy and other terms of employment *How diversity and capabilities are sustained and developed* Positively encourages diversity Values individual capabilities Reviews the effectiveness of training Promotes continuous learning *How the organisation agrees targets and reviews performance on the culture of respect* Ensures that personal objectives are aligned with showing respect to others Appraises and helps people to improve their behaviours towards others *How the organisation manages partnerships and resources* Provides access to relevant information to employees and other stakeholders Protects the confidentiality of personal information *How the organisation identifies processes which are key to a culture of respect* Defines key processes Evaluates the impact of key processes on the culture of respect *How processes are reviewed and improved* Identifies how processes can be improved Uses data from benchmarking, focus groups, etc., to establish best practice Identifies and agrees stretching targets		

Figure 9.4 (continued)

Criteria	Score*	Comments
How processes are changed and benefits are evaluated Agrees appropriate methods of implementing change Communicates process changes Reviews process changes to ensure improved results are achieved *How the satisfaction of employees is measured* Use of surveys, structured appraisal and focus groups Identify levels of satisfaction *How the culture of respect impacts on society* Use of surveys, structured appraisal or focus groups Involvement in benchmarking Accolades and awards received Recognition by external bodies (EFQM, IIP)		

* Scoring on a 10-point scale where: 10 – evidence of a standard of excellence that could not be improved, 0 – no evidence of any achievement on this measure.

initiatives were found to be particularly successful (Crosby, 1989). Recognition is particularly important when attempting to achieve changes in culture (Williams et al., 1993). Recognition does not have to be financial to be effective, indeed a financial reward can be counter-productive. A number of organisations have used the formal presentation of certificates or medals for excellence as an effective sign of recognition (Bank, 1992). Recognition can also be achieved in the appraisal process where appropriate objectives are set to assess the levels of respect shown to colleagues and teams.

European Foundation for Quality Management (EFQM)

The EFQM was formed in 1988. The foundation recognised the potential for competitive advantage through the application of total quality management in organisations. The EFQM, with the support of the European Organisation for Quality and the European Commission, took a leading role in establishing the European Quality Award in recognition of the achievement of organisations in pursuing total quality. While the award provides a high profile for total quality, its importance is more in the self-appraisal process associated with determining

Figure 9.5 The business excellence model (© 1999 EFQM)

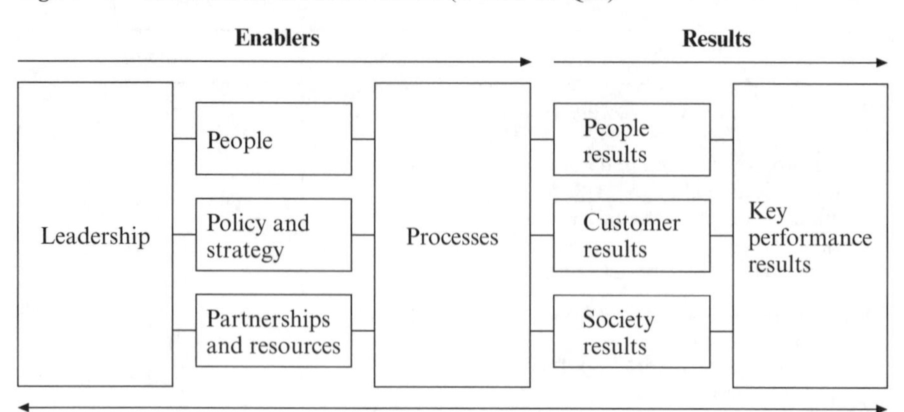

how organisations are progressing in total quality, which involves a regular and systematic review of the organisations' activities and results. The process allows organisations to identify their strengths and the areas where improvements can take place. The EFQM model, which was revised in 1999, is illustrated in Figure 9.5.

The business excellence model consists of nine criteria grouped in two main areas: *enablers* – how we do things – and *results* – what we target, measure and achieve. The model also reflects the importance of innovation and learning in achieving competitive advantage. Each criterion has a high level definition and is supported by a number of subcriteria. The subcriteria pose questions to be considered when assessing the progress of an organisation towards excellence.

Investors in People (IIP)

Investors in People is a national standard which sets the level of good practice for improving an organisation's performance through its people. The standard provides the basis for continuous improvement of both the organisation and its people. Although it is not directly linked to total quality, it does provide a framework that recognises the importance of people in achieving organisational goals. Investors in People is based on four key principles:

- Commitment to invest in people to achieve business goals.
- Planning how individuals and teams are to be developed to achieve these goals.
- Action to develop and use necessary skills in a well-defined and continuing programme directly tied to business objectives.
- Evaluation through measuring progress towards goals, value achieved and future needs.

The four principles are supported by 23 indicators that tell the organisation whether the national standard is being met.

Although not specifically designed to assist in the development of a culture of respect the EFQM model and IIP assessment and monitoring processes can both be used as tools to bring about the necessary cultural changes in organisations.

The challenge for organisations

This chapter has illustrated how organisations can use tried and tested tools and techniques to develop an organisational culture that is built on respecting the dignity of employees. The continuous improvement approach appears to be one that is particularly suited to organisations where the employees themselves identify the standards of behaviour which are necessary to demonstrate a culture of respect. This approach moves away from the more prescriptive approach in which standards are applied irrespective of the particular climate or needs of the organisation and its employees. Whether the organisation chooses to use the EFQM approach or prefers the IIP assessment procedures is not important. Both systems enable the organisation to access a structure of support with the objective of creating a culture which will not only be more respectful of the needs of employees but also contribute to the business itself through the changes in attitudes and motivation of its workforce.

References

Bank, J. (1992) *The Essence of Total Quality Management*, London: Prentice Hall.

Bertram, D. (1991) 'Getting started in total quality management', *Total Quality Management*, 2, 3, 270–82.

Clarkson, P. (1995) *Change in Organisations*, London: Whurr Publishing.

Collard, R. (1993) *Total Quality: Success through People*, London: IPM Publishing.

Crosby, P. (1989) *Let's Talk Quality: 96 Questions That You Always Wanted to Ask Phil Crosby*, New York: McGraw-Hill.

Debell, B. (1997) *Conciliation and Mediation in the NHS: A Practical Guide*, Abingdon: Radcliffe Medical Press.

Deming, W. E. (1982) *Quality, Productivity, and Competitive Position*, Cambridge, MA: MIT Centre for Advanced Engineering Study,

EFQM Excellence Model (1999) The British Quality Foundation, 32–4 Great Peter Street, London SW1P 2QX.

Easton, G. S. (1993) 'The 1993 state of US total quality management', *Californian Management Review*, 35, 3, 279–82.

Ghoshal, S. and Bartlett, C. A. (1997) *The Individualised Corporation: A Fundamentally New Approach to Management*, New York: Harper Business.

Haksever, C. (1996) 'Total quality management in the small business environment', *Business Horizons*, March/April, 33–40.

Joyce, M. E. (1995) *How to Lead Your Business Beyond TQM: Making World Class Performance a Reality*, London: Pitman.

Kano, N. (1993) 'A perspective on quality activities in American firms', *Californian Management Review*, 35, 3, 12–31.

Manganelli, R. L. and Klein, M. M. (1994) *The Re-engineering Handbook: A Step by Step Guide to Business Transformation*, New York: American Management Association.

Manz, C. C. and Sims, H. P. (1993) *Business without Bosses*, New York: Wiley.

Mossink, J. (1997) *Instruments and Models to Assess Costs and Benefits at Company Level*, Amsterdam: NIA TNO B. V.

Oppenheim, A. N. (1966) *Questionnaire Design and Attitude Measurement*, London: Heinemann.

Oxenburgh (1997) 'The productivity model: a cost-benefit computer model for implementing health and safety at the workplace' in J. Mossink and F. Licher (eds) 'Costs and benefits of occupational health and safety', proceedings of the European conference on costs and benefits of occupational health and safety, The Hague, 28–30 May 1997, Amsterdam: NIA TNO B.V., 361–5.

Prokesch, S. E. (1997) 'Unleashing the power of learning', an interview with British Petroleum's John Brown, *Harvard Business Review* (September–October).

Reynolds, C. (1997) 'Beyond dispute', *People Management*, 3, 23, 40–2.

Tehrani, N. (1996) 'The psychology of harassment', *Counselling Psychology Quarterly*, 9, 101–117.

Thiagarajan, T. and Zairi, M. (1997a) 'A review of total quality management in practice: understanding the fundamentals through examples of best practice applications – part 1', *The TQM Magazine*, 9, 4, 270–86.

Thiagarajan, T. and Zairi, M. (1997b) 'A review of total quality management in practice: understanding the fundamentals through examples of best practice applications – part 2', *The TQM Magazine*, 9, 5, 344–56.

Williams, A., Dodson, P. and Walters, M. (1993) *Changing Culture*, 2nd edition, London: Institute of Personnel Management.

CHAPTER TEN

Don't suffer in silence

Building an effective response to bullying at work

STEVE RAINS

Introduction

In the early 1990s few organisations had recognised the extent of bullying taking place in the workplace. While the personnel managers may be faced with dealing with the occasional serious harassment case, few organisations had considered the need to develop formal policies and procedures to deal with the problem of bullying. The research of the journalist Andrea Adams (1992) raised awareness in organisations of the issues involved and this influenced a number of organisations to take a closer look at whether bullying was a problem within their own organisation.

The Post Office, as one of Britain's largest employers, was particularly keen to ensure that it had appropriate policies, procedures and sources of support to address bullying. The problem that faced organisations at this time was the absence of information and advice on the best way to achieve an organisational environment where bullying and harassment were rejected as ways of behaving towards fellow employees. This chapter will look at the background and introduction of a practical 'listeners' scheme that was introduced into a Royal Mail division to encourage and promote a cultural change which had the aim of eradicating bullying and harassment within the workforce. The structure and management of the listeners scheme will be described, together with the way in which listeners were selected and trained. There will also be a description of how the organisation monitored the quality of the support and its impact on the workforce. Finally, the chapter will discuss what has been learned and will offer advice to other organisations wishing to develop similar schemes.

Looking for best practice in organisational responses to bullying

At the end of 1992 the Royal Mail's south central (RMSC) division undertook research to identify the reasons for the problems within its workforce. There was a particular concern, which was the difficulty experienced in persuading women to apply for promotion in the sorting offices. Feedback from discussion groups and from employee opinion surveys began to reveal a fairly consistent pattern of bullying and harassment. This pattern included inappropriate touching, insults, jokes and suggestive remarks. A consistent comment was that there was no point in complaining about being bullied as it would only make matters worse (RMSC, 1992a). It became clear that there was little confidence in the formal complaints procedures and that the division was seen as failing in its responsibility to deal with bullying and harassment effectively. In common with other organisations, this lack of effective action had led to employees putting up with the bullying behaviours in the belief that complaints would not be taken seriously (Rayner, 1994). What was clear from the research was that, unless the organisation could provide someone for the employees to talk to in confidence – a person who could be trusted, who was not part of the formal complaints procedure and who understood the environment on the sorting office floor – then the situation was unlikely to improve.

In an attempt to find a way to deal with these problems, the then employee relations manager decided on a course of action which involved identifying organisations against whom he could benchmark with the aim of introducing the best practice into the division. Unfortunately, in 1992, although a number of organisations were considering a variety of programmes to deal with the effect of bullying and harassment, none had actually implemented their ideas. This faced RMSC with two possible options; firstly, it could wait until another organisation had trialled their scheme and then evaluate and adopt that scheme to the needs of the Royal Mail. Or, secondly it could go ahead and create its own scheme that would be designed to meet the specific requirements of the Royal Mail. Unsurprisingly, RMSC chose the second option. It is perhaps worthy of note that by 1994 RMSC had already gained a reputation for its listeners scheme and was being regularly contacted by other organisations that had heard of it. These organisations wanted to benchmark with RMSC before introducing similar schemes into their own organisations (RMSC, 1994).

The size of the problem

In the early 1990s the Royal Mail was recognised as one of Britain's leading employers, with well-developed and sound policies on equal opportunities and harassment. The Royal Mail had trained its managers in the use of these equal opportunities policies and breaches of the policies were generally dealt with appropriately. As with many other organisations the relatively small number of harassment cases that came to the attention of the senior managers were generally interpreted as a sign that all was well and that bullying and harassment were not a problem (IRS, 1999). However, those cases that were being seen were just the tip of an iceberg, one-off outrages or a final straw in a chronic case of bullying or harassment. With this apparent lack of evidence of a problem it was easy for senior managers to become complacent. However if, as the saying goes, 'there is no smoke without fire' is it also true that the converse also applies in that 'there is no fire without smoke!'? If the problem is not visible, then there cannot be a problem.

What RMSC came to realise was that it had been dealing with the glowing embers of bullying and harassment that, despite being hidden under the surface, were responsible for the uncontrolled visible flare-ups of some of the more serious cases. Evidence to support this view was soon discovered in the statistics obtained from an employee opinion survey undertaken in 1992. Although very few cases of bullying or harassment had been reported, employees indicated that bullying and harassment was taking place. The problem appeared to be the lack of confidence in the way the organisation handled bullying and harassment with only 47 per cent of employees believing that the Royal Mail took bullying and harassment seriously (RMSC, 1992b).

Having gained evidence on the nature of the problem, RMSC wanted to establish why this problem had arisen. A number of focus groups were held with representatives of all groups of staff. From the focus groups a clear pattern of responses emerged that suggested the most important issue was the unshakeable belief that there was no management commitment to treat bullying and harassment seriously. The problem was compounded by feedback from frontline managers who had formed the view that it was easier for them to sweep bullying and harassment problems under the carpet as the existing processes to deal with it involved the daunting task of going through the formal complaints procedure. The awareness and use of informal processes for handling bullying and harassment did not appear to be an option as most employees and managers at that time believed that, if bullying or harassment was reported, then the manager or union would be duty bound to take up the issue formally. For many employees who were experiencing the inappropriate behaviour of a fellow worker the thought of taking out a formal complaint was seen as inappropriate. However there appeared to be little guidance on what else might be available to resolve the issue. Although informal processes were written into the equal opportunities procedures there was little guidance on the workings of the informal approach. What the employees wanted was someone to talk to about what was happening and who could give simple information on the options that were open to them to resolve their difficulties.

A peer listening scheme

RMSC decided that the most appropriate approach would be to introduce a peer listener scheme. The peer listeners would listen to their colleague and provide information on the choices that were available to resolve their difficulties. The peer listeners could also help their colleague by providing support in the handling of formal procedures, should that be the chosen route. The scheme aimed to ensure that all employees could contact a peer listener at any time in the knowledge that what they said would be kept in complete confidence. The peer listener and colleague would discuss the nature of the problem over the phone and, where more appropriate, meet face to face either at work or at a place outside the place of work. Having listened to their colleague's story sympathetically but impartially, the peer listener would be able to advise the employee in a number of areas including:

- Which procedures were available to be used to deal with the bullying and harassment.
- The potential role of the peer listener as a supporter through the formal or informal procedure.
- The peer listener's considered view on the situation.
- The range of professional help which was available from the Royal Mail.
- Any other issues related to bullying and harassment policy and procedures.

What the peer listeners would not do was to make a decision on whether or not the employee should take their complaint of being bullied or harassed further.

RMSC took the decision to call the peer listener volunteers 'listeners'. This choice was to avoid any misunderstanding about the nature of the listener role and to maintain a clear distinction between these peer volunteers and the professional counselling service offered by The Post Office's Employee Health Service that provided well-being advice and counselling support for all Royal Mail's employees. Indeed, the listener's role was to act as a bridge to the professional help that was available from the Employee Health Services.

Recruiting the listeners

Having established the need for a scheme, RMSC began the process of defining the specifications for the scheme and establishing the qualities and competencies required by a listener. It was decided that the listeners would be unpaid, but well supported, volunteers. This decision was not as a result of any reluctance to invest resources into the scheme, on the contrary both the Divisional Board and Royal Mail National Headquarters had given the scheme their full support. The key reason was that if the listeners were paid there might be an encouragement for individuals to apply to do the work primarily for the money. The scheme needed the right people to undertake this demanding work for the right reasons.

It was decided to trial the listener scheme in the Oxford and Hemel Hempstead postcode areas within the division. A job description was written and an advertisement for volunteers was put on the noticeboards of all the offices in the two areas. The advertisement sought employees of the highest calibre to fulfil the role of listener. Applications were invited from all the employees within RMSC with the exception of senior managers, personnel managers and union representatives, the reason for their exclusion being that these groups were already involved in the formal investigation process and that their listener role would conflict with their primary role. The advertisement specifically encouraged applications from postmen and postwomen. It enabled a degree of self de-selection by stressing the demanding nature of the role, the personal qualities needed for it, and that there would be no additional pay for listeners. There was some concern at that time that the level of commitment demanded by the listener role, which involved giving up both personal time and time at work, might reduce the numbers of suitable applicants; however this concern was unfounded as the level of interest in the role was high.

The selection process

The personal qualities emphasised were those crucial to the role – excellent interpersonal skills, the ability to listen, self-motivation, impartiality, resilience and of course the most important quality of respecting confidentiality, no matter how strong the pressure to break it.

Twenty-three applications were received; this allowed us to aim for four listeners in each postcode area. This number provided the employees in each area a choice of listener. This was important as some people may wish to see a colleague in their own office while others may deliberately decide to see someone from another area. Depending on the type of case, there may be a preference for a listener to be either a male or female colleague. The decision was taken to issue photo posters of the listeners as soon as they were appointed and allow employees to choose their own listener. The number of listeners in each area also provided sufficient cover to spread the workload and to keep the task manageable.

The selection process was then undertaken. This was the first of two pass/fail stages to becoming a qualified listener. In the selection process the candidates were put through an interview based on the personnel assessment used in the Royal Mail managerial selection processes. This selection process tested purposefulness, resilience, interpersonal skills and maturity. Qualified assessors using these selection criteria explored the candidate's past experience and actual responses to situations in order to gain positive or negative evidence for each quality. The final stage of the selection process was a role-play exercise. The role-play used an example case to test the candidate's attitude to confidentiality and integrity. Adopting an alleged victim as a client and then subsequently (without the knowledge of the 'victim') being asked for advice by the alleged harasser who naturally told a very different story. This final exercise was particularly useful. Candidates eager to please respond by either being willing to take both the victim and bully on as clients or, worse, using information from one to advise the other – needless to say candidates who responded in these ways failed the selection. This exercise was, however, realistic. Listeners when appointed would have to resist approaches for information from managers or indeed union representatives who did not fully understand the process. At this stage in the process of introducing the listener scheme both senior management and the union had as collective bodies, signed up to support this initiative.

From those candidates who passed the assessment, eight were to embark on the next phase. The next stage consisted of a pass/fail training course specially designed and delivered by the Post Office Training and Development Group at the Management Training College in Rugby.

Listener training

The week-long course concentrated on the necessary skills and knowledge required for a listener. The key areas were:

- the Royal Mail's conduct code procedure;
- the Royal Mail's harassment procedure;
- defining harassment/bullying – its cause and effects;
- outlining the role and responsibilities of a listener;
- defining and practising listening skills (using CCTV).

The week ended with a pass/fail test involving a written test which examined the candidate's knowledge of harassment and bullying and the Royal Mail policy, and a practical session in which they were required to demonstrate their use of key listening skills.

It was made clear at the outset of the training that if this final hurdle were not cleared, the candidate would not become a listener. However all eight candidates passed the assessment which was a testimony to the first phase selection assessment. It was not until the fourth course that any of the candidates were assessed as not competent at this stage of the training.

Implementation

Having previously 'sold' the scheme to the unions and senior management, the next task was to prepare the ground for the new team before they 'went live'.

The purpose of this preparation was twofold. Firstly it was to publicise the availability of the listeners to help fellow employees; secondly it was to dispel any fears from managers, union representative and employees. It was also important that managers were asked to assist by ensuring that time and facilities were made available. Reassurance was given to managers and union representatives that the listeners were not going to take over their roles in any of the existing processes; indeed it was emphasised that the listener's involvement was designed to be of help. However it was necessary to explain that this help would fall short of sharing any personal information or confidential material. This reassurance was important in helping to show employees that the listeners were truly independent, totally confidential channels for them to use. The key issue for the client was that at all times it was them who remained in control. The listener might offer advice and steer them in the right direction in terms of procedure or to the availability of expert help – but it was and was *always* the client's decision on whether or not to proceed. It was found that some employees simply felt better by downloading their concerns to a sympathetic ear and, if that is all they require, then that is their prerogative.

All employees received a letter introducing and explaining the listener scheme. A full colour laminated photo poster, displayed in each unit, supported this letter. Both the letters and the posters gave the names and telephone numbers of the listeners and the 24-hour harassment hotline telephone that was managed by the equal opportunities manager. The posters went under the strapline 'Don't suffer in silence'. Finally, the scheme was publicised in both the divisional and area internal magazines.

Response

RMSC was delighted with the positive response to a trial that was an unqualified success. Particularly pleasing was the fact that the careful preparation, selection and training paid off. Cases were dealt with efficiently and without any problems for either the listeners or their clients.

Today, there are still listeners in the area previously covered by the division (eleven postcode areas) in Berkshire, Oxfordshire, Wiltshire, Hampshire, Surrey, Buckinghamshire and outer London.

Initially, the introduction of the scheme led to an increase in the number of cases which were reported formally, in addition there were a number of cases that were resolved without the need for recourse to formal action. This initial increase was to be expected; it confirmed the belief that RMSC had been able to uncover a previously hidden problem. A 12-month period resulted in 178 contacts to listeners with about 5 per cent of these proceeding to formal action with some unfortunately progressing to serious disciplinary hearings. It is, however, a measure of success for the scheme that an increasing proportion of cases have been resolved informally and that this is a measure of the way I believe the scheme has improved the culture within our area.

The listeners have, on a number of occasions, acted as an important bridge to the Employee Health Service. The listeners have become recognised as well trained experts on our bullying and harassment procedures to such an extent that some managers have taken the opportunity to seek listener advice when confronted with a problem. Another pleasing statistic comes from the employee opinion survey. This survey shows that from the original starting point of 47 per cent, the number of employees who currently believe that RMSC takes harassment and bullying issues seriously has risen to 77 per cent. The number of contacts with the listeners and employee reliance upon listeners has fallen slightly from the peak figures reached after the first three years of the scheme. It is believed that this is the indication that the important change in culture has started to become established. It has also been pleasing to see that the scheme has acted as a career development opportunity with two members of the original listener team having moved on to careers in professional counselling.

Support

An important issue that cannot be overstressed is the need for support and refresher training for the listeners. Support can be looked at in two ways. Firstly, in terms of physical support for the listeners with respect to the time and equipment required for undertaking the role. This support need not be excessive, indeed it was found that some listeners had to be encouraged to reclaim expenses which gives some indication of the level of their dedication.

Secondly, moral and practical support is perhaps even more important. In RMSC this support was provided by the Equal Opportunities manager, David Vaughan, who undertook careful monitoring of the team. David proved to be a genuine, enthusiastic and keen professional and, when he was promoted, Bob Huckerby proved to be an extremely able replacement.

We were also given professional sessions with our Employee Support Team, at the time led by Noreen Tehrani. The Employee Support Service has since been absorbed within the Employee Health Service which now provides this support.

Finally, we provide an annual listeners conference that takes place over two days and serves two purposes. The first is an opportunity to reinforce and update training. The conference is well supported by Anita Thomas from the Training and Development Group at Coton House in Rugby. Anita has been involved with the scheme since its inception. The second is that it acts as an opportunity for the reward and recognition of a dedicated and hardworking team of volunteers who have made a real difference to the lives of the 16,000 employees who work within the area covered by the division.

I would recommend the listener approach to any organisation; it certainly has worked for RMSC. Here are the key points in summary:

- Be selective – only use the right people for the sake of the clients and the listeners themselves.
- Expect an increase in cases and discussion of the problems at the beginning of the scheme.
- Provide the support and facilities to meet the physical and psychological needs of the listeners.

I have shown that a listener scheme can improve your culture and the attractiveness of your workplace. It does, however, require the investment of time and resources, but surely it is worth it to ensure that your employees do not 'suffer in silence'. Good luck!

References

Adams, A. (1992) *Bullying at Work: How to Confront and Overcome It*, London: Virago Press.

IRS (1999) 'Bullying at work: a survey of 157 employers', *Employee Health Bulletin*, 8, 677, 4–20.

RMSC (1992a) 'Harassment: is there a problem?' *The Listener*, 1, Winter 1992, Royal Mail South Central.

RMSC (1992b) 'Employee opinion survey', internal survey.

RMSC (1994) 'Next steps', *The Listener*, 3, Spring 1994, Royal Mail South Central.

CHAPTER ELEVEN

Issues for counsellors and supporters

BRIGID PROCTOR AND NOREEN TEHRANI

Introduction

This chapter looks at the need to provide practical and emotional support for people in conflict and describes the sources of the support that are available. Working with bullies and the bullied can be both difficult and exhausting and even the most resilient and resourceful professionals may find that they lack the knowledge or skills to deal with particular situations, people or organisations. The chapter explores the quality of communication which is essential for effective counselling, advising and guiding both the bully and the bullied. It begins by looking at the organisation.

Organisations are complex systems with their own rules and cultures. The way that organisations manage people plays an important role in the creation of a culture that is either respectful or bullying towards its employees. The organisation's legal duties of care frequently require very different working contracts for the counsellor or supporters than would normally be the case. This means that anyone who chooses to undertake this work should be very clear about the impact that this organisational dimension might have on the nature of the relationship that they can establish with the many clients, customers and stakeholders involved. It ends by addressing the need to protect supporters and counsellors from the impact of being involved in this kind of work and suggests ways in which supporters can help themselves and how organisations can discharge their responsibility to support supporters.

Working with the organisation

It is perhaps worthwhile looking at the historical context within which advice, support and counselling for employees has developed within organisations. The first employee welfare services were created in 1886 when the Rowntree factory in York appointed a welfare officer. The idea that organisations should take an interest in the well-being of their workforce gained support and, by 1913, the Welfare Workers Association had been formed (Edmonds, 1991). This association was the forerunner of the present day Chartered Institute of Personnel and Development. These early personnel and welfare professionals faced a dilemma – 'where is my focus, the organisation or the employee?'

This conflict continued until the mid-1960s when a clear distinction was made between the personnel professional who chose to act as an arm of management involved in improving business efficiency and the welfare officer whose role was to look after the personal and social needs of employees (Fox, 1966). For many professionals this separation resolved the difficult problem of how to handle the competing demands of organisation and employee. However this simplistic solution failed to recognise that these complex dynamics were at the heart of many of the difficulties faced by both employee and organisation (Butler, 1999).

Research work in identifying the most effective solution for dealing with issues such as bullying recognises the needs and sensitivities of both the individual worker and the organisation (Tehrani, 1996). Many organisations choose counselling as a best way to respond to the bully and the bullied. However there is evidence to show that organisations are uncertain of what counselling entails and there are few attempts by organisations to define or monitor what is provided (Berridge et al., 1997). An individualistic approach that focuses exclusively on the needs of the employee and fails to recognise where the cause of the conflict is within the organisation can result in an increased potential for the problem to be perpetuated. Many counsellors, advisers or supporters regard client autonomy and self-determination as central to their professional practice (Feltham, 1995). Any support system which maintains a separation between employee and the organisational context in which they exist runs the risk of becoming part of the problem rather than its solution (Lane, 1990).

The source of support

Employees experiencing bullying at work are faced with a number of potential sources of support. Chapters 8 and 10 provide examples of two successful approaches. However there are other sources of support. Table 11.1 illustrates some of the ways that employees may be helped.

It is important to recognise the wide range of people that may become involved in providing support for employees exposed to bullying. While it is important for employees to be aware of the different sources of support it is also important for organisations to be aware of the number of people who may be indirectly impacted by bullying.

Table 11.1 Sources of support for employees

Source of support	What is provided?	Provider?
Telephone helpline	These may be specific to bullying or harassment or part of a counselling service. They may offer information or may also act as a route to a counselling/support service.	Union, internal or external counsellors/ advisers
Information/advice	Information on the organisational policy and procedures for dealing with bullying.	HR department, union, line managers
Confidential supporter	Someone to talk to who can provide information and support which allows the employee to decide on what they do.	Trained peers, counsellors or welfare officers
Formal process supporter	Information, support and guidance related to the formal complaints procedure.	Colleague, peer, HR department, union
Training and education	Seminars and training sessions on bullying and harassment. This may include looking at the organisation's culture, policies, procedures and the psychological impact of bullying.	HR department, specialist trainers, union, managers
Mediator/conciliator	Non-formal approach to resolving interpersonal problems.	Trained mediator or conciliator
Counsellor	An approach to dealing with the personal and emotional impact of bullying and harassment.	Trained counsellor

Organisational roles and contracts

One of the important aspects of providing an effective service to support organisations and employees when dealing with bullying is the assignment of rights, roles and responsibilities. The development of effective and appropriate contracts with the organisation and its employees is the best way to avoid being faced with conflicting obligations (Bond, 1994). The contract with the organisation needs to recognise that the organisation itself is a client, with sensitivities and vulnerabilities, and deserves to be treated with respect. What kinds of contracts should be formed between a provider of counselling or support and the organisation? The explicit contract should contain a clear description of the expectations of the organisation and the service provider with regard to the support, facilities, service range, standards of performance, practitioner

Table 11.2 Organisational expectations of counsellors and supporters

Positive expectations (enablers)	Negative expectations (detractors)
Takes a global rather than narrow view	Is always on the side of the client
Understands the needs of the business	Is only interested in the needs of the client
Is prepared to work with the manager	Believes that the manager is the problem
Inquisitorial approach to problem resolution	Adversarial approach to problem resolution

qualifications, mechanisms for monitoring and auditing, routes for reporting and fees (Bull, 1995). It is recognised that few counsellors and supporters have the benefit of comprehensive contracts setting out the responsibilities of all the parties, although some organisations have moved towards this position (see Chapter 10). As a minimum a contract should describe:

- the nature of the work to be undertaken;
- the context and situation in which it is undertaken;
- the resources which are to be made available;
- the reporting and evaluating processes.

Equally important is an awareness of the implicit or psychological contracts which are usually unwritten and are made up by a number of assumptions built up over time (Rousseau, 1995). Much of the literature on psychological contracts has focused on the contract from the point of view of the employee or service provider (Schein, 1980). Clearly contracts are two-way and organisations will have unwritten expectations of how a counsellor or supporter dealing with bullying should behave in relationship to the organisation. Of greatest importance is the need to support organisations that are trying to achieve a culture of respect. Such support can be communicated in the same way as when working with an individual. The organisation's successes should be recognised and rewarded. Where there are failures, the organisation should be encouraged to acknowledge the failure, to identify why it occurred and then be helped to make the necessary adjustments. By adopting this approach which does not involve shaming or denouncing the organisation publicly, the counsellor can have a much greater impact on the way the organisation grows and develops. Table 11.2 gives a number of positive and negative organisational expectations of counsellors that can either help or hinder the development of a culture of respect.

The expectations, which are created between organisations and counsellors, are interactional. For example if the representatives of the organisation have negative expectations of the service provider, it is likely that these expectations will influence the way in which the counsellor or supporter approaches the work.

Where the expectation is that the counsellor will take sides it is likely to be difficult to build up the atmosphere of trust and respect essential for a more balanced approach. On the other hand, organisations anticipating a working relationship based on mutual trust and respect are likely to find their counsellors and supporters responding in a similar fashion. Seeking to have respect and understanding for the needs of the organisation is therefore the necessary background to becoming involved in supporting or helping those who complain of being bullied and working with those who are using bullying behaviour.

Working contract and boundaries

Creating clear agreements with any party concerned signals acceptance and respect. This means that the helper communicates the roles and responsibilities that he is offering to undertake and the overall purpose of the work that the parties can do together. This definition of roles, responsibilities and tasks is preset as part of the helper's contract with the organisation. It will also make clear what is not on the agenda – the boundaries of rights and responsibilities. Communicating means more than telling or announcing and we will return to a reminder of the skills of contracting and negotiating later. The clarity and honesty of the working agreement will determine how safe the environment is and how possible it will be for a trusting working relationship to develop. The nature of the working contract will depend on how the helper, or the helper in conjunction with the client, decides to work.

Skills for contracting and negotiating

The skills for negotiating working agreements and making clear what is or is not negotiable are an essential part of a helper's repertoire. These skills are essential in the establishment of a contract with the organisation, but are equally important when setting up a working contract with a client. One of the main skills when working with clients is the ability to distinguish between purpose stating and preference stating. Purpose stating makes it clear to the client what can and cannot happen in any given situation. This is determined by the nature of the contract with the organisation and with professional codes of ethics. Examples of the purpose of the working contract might include the number of sessions that can be provided and any limitations to the role of supporter (see Chapter 10). The supporter needs to convey where arrangements are non-negotiable.

On the other hand, preference-stating makes clear the helper's preferences – 'I think it would be useful for me to talk with your manager. What do you think?' It is important that supporters do not present a hidden non-negotiable agenda as an option as this results in confusion and undermines trust. Before asking for the client's preference or opinion, supporters need to consider whether the option is really available and if, when the choice is made, the decision will be supported.

Clients frequently fail to 'take in' carefully laid out contracts, particularly when they are anxious, distressed or preoccupied. Therefore it is useful for the supporter to have written information that can be read ahead of an interview and referred to again afterwards. There are facts that have to be made clear and understood before starting to work with the client and other things about the work that can only be understood after the client has gained some experience. The helper needs to bear in mind that, by definition, a contract or working agreement has to be understood and 'agreed' by both parties. Helpers need to keep the contract as a backdrop to the work and point to the meaning of roles, responsibilities and boundaries when what is being said in the course of work illustrates the meaning of the contract. For example when the advantage of contacting the client's manager is recognised the helper might say, 'Do you remember that when we first met I said that there may be times when it would help to speak with your manager? I also said that this would only be with your permission and that we would agree what could be said.'

Supporter, counsellor, conciliator or advocate?

The helper will probably first work with the clients in the limited role of supporter/adviser, listening and informing clients of their rights. However, this may lead to offering to take the role of counsellor – seeking to listen, understand, clarify, support and inform the client so that the client can make choices and be enabled to cope resourcefully with the situation. These are different roles and tasks that require the helper and client to be clear of the role and task in which they are engaged.

In addition, at some stage, the helper may have to decide whether to undertake a more complex task – that of working with more than one party. If this offer is made, the contract needs to be particularly clearly agreed. Will the role be that of conciliator who is equally open to both sides of the story, helping the parties come to some shared solution? Or will it involve approaching a second party in the role of advocate for the original client? Such agreements raise ethical considerations that need to be carefully thought through before engaging with a second party. Some of the problems include:

- What will be confidential?
- How will confidentiality be ensured?
- If the helper agrees to conciliate, who is 'the client'?
- What will be the ground rules for acceptable behaviour if both parties are seen together?
- What action will the helper be 'licensed' to take if the ground rules are not adhered to?
- What constitutes 'informed consent' to conciliation?

Case study

Two employees agreed to try conciliation as a way of resolving a dispute between them. The terms of the conciliation were agreed between the counsellor, the organisation and both employees. The first stage of the process involved each employee going through their story in fine detail with the counsellor, producing a summary of the story to check that her understanding of what had happened was correct. On the second meeting one of the employees decided not to pursue conciliation: 'I feel so angry about what happened to me I do not believe that I can ever work with her again.' The counsellor was then faced with a situation where all parties had agreed to a particular approach to resolving a difficulty but now one person wanted to renegotiate a different approach. The counsellor who had been committed to the conciliation felt confused as there were many unanswered questions. Did I not explain the process well enough? Was there a hidden agenda? Is it appropriate to continue to see both employees? What will the organisation think?

Supporter skills

The range of skills required by supporters depends on the range of helping tasks and roles that they may be called on to undertake. Equally, the roles they undertake should, ideally, be limited to the skills they currently have. At the very least, they should be aware of the skill required by some of the more complex helping tasks, and do their best to identify their areas of strength and those they need help and support to develop.

All the helping tasks require the ability to lay aside moral judgement, and to listen with close attention, not only to the story but also to the person who is living the story. The listening may have different purposes at different stages of different overall tasks.

Examples would be listening for:

- ways to develop an empathic understanding;
- unexpressed emotions or thoughts which may inhibit the client from appreciating his situation fully;
- evidence of untapped resources;
- evidence of misinformation or lack of information;
- possible realistic action plans;
- clues on the best way to work in the total situation.

Communicating with the client requires helpers to appreciate and understand the need to test that their understanding is correct. The communication process entails paraphrasing and summarising what has been heard. 'Getting it wrong' (initially or from time to time) may be as valuable as 'getting it right', particularly when it enables the client to clarify their meaning. These basic counselling skills should be enhanced by the ability to change gear – to aid the client's and your own understanding and appreciation of the situation in a fuller and less stereotypical way. 'Oh, that's what it is like' or 'I couldn't understand why, but now . . .' This happens through:

- helping the client listen more carefully to him or herself;
- noticing and tentatively offering connections, missed pieces of the jigsaw, hypotheses;
- 'playing-back' what has been said in a way that adds fullness and meaning to that which has been hinted at, but not stated.

Other important skills are those for helping people solve problems, set goals and monitor their own thoughts, emotions and behaviour in the light of what they value and want for themselves. These skills require the helper to be both realistic and optimistic. They also require temperance and self-discipline – it is the client's life and situation. Clients may appreciate help, and ask for advice, but if the goals set by the helper are not their own, the resources required to achieve the goal are not available or the solutions are not consistent with the client's value system, then the client will not be helped.

Case study

A woman who had been bullied by her manager wanted the bullying to stop but felt afraid of what might happen if she formally reported the problem. The helper described how the woman might deal with the situation and how she would be protected. Regardless of the assurances and the support that were made available, the woman was still afraid to go ahead but equally was not content to accept the existing situation. The helper, almost in despair, asked the woman if she knew if the manager had bullied any of her colleagues. The woman thought for a while and said that she did not know but that there was one person who always seemed to avoid him. The woman then said that if she was not the only person who had been bullied then she would not feel so isolated. The whole focus of the work then moved to helping her find a way of talking to the colleague and, subsequently, to exploring how they both could make their complaints.

This helper had not been unhelpful, but initially the approach had not been directly useful to the client arriving at her own solutions. However by inviting the client to explore the situation more widely, the helper had enabled the client to find her own way to deal with the situation. Hearing a solution that they do *not* want or cannot use helps many people. What is important is for the helper to recognise when the client is finding his or her own solutions and not to get locked into the need to come up with their own 'right solutions'.

Team work

Helpers usually work as part of a team. Not only do they have to be clear about their role and responsibilities within the team but also they have to remember the roles and responsibilities of other team members such as the line manager and union representative. Helpers may need to remind colleagues about the contractual ground-rules on access, confidentiality and representation. However, if the helpers are to gear the resources of the organisation to the needs of the client, they will have to bear in mind the difficulties, frustrations and limitations of other team members. In the following example the helper was working with a number of different groups, all of whom had a role to play in addressing bullying at work. The key players were the abused employee, her manager and colleagues, human resources, occupational health and the organisation's own security department. The helper was aware of some of the issues that faced the other players but tried to maintain her focus on the employee.

Case study

An employee had been the victim of a series of unpleasant and demeaning anonymous notes at work. Despite a considerable effort on behalf of the organisation, the culprit was never identified. The employee was so upset by the content and the number of notes she went off sick. The helper invested a lot of effort in trying to identify what could be done to help the employee to return to her job and was actively involved in working on a rehabilitation programme with the occupational health adviser. On the day of her return to work graffiti was found which totally unnerved the employee. She did not feel safe. Despite the efforts of the security and personnel departments, no one could be identified as having written the notes. The helper felt there was little that she could do to resolve the situation and began to wonder whether the investigation had been as thorough as it should have been.

The helper would have found it easy to see the organisation as the persecutor of the employee and had to continually remind herself of how easy it was to become trapped in the drama of the situation and lose her true role in working towards a solution.

Helping in the service of a culture of respect

Counsellors and supporters have an inherent difficulty in their role and task. On the one hand, they are a place of refuge, sympathetic listeners and supporters. On the other hand, they are committed to being a holder of the bigger picture; someone who can have alternative perspectives on how the person who is experiencing bullying can manage the choices that are realistically available. As has already been mentioned, they may have a role in working with the person complained against in order to raise awareness of the effect of the behaviour experienced as bullying. This task requires a commitment to an honest appraisal of the situation. It is useful to consider some of the basic values that underlie the model of helping that is proposed and identifies some of the traps in which unwary supporters, with the best intentions, can find themselves ensnared.

Truth, self-regard and choice

In a seminal work in which he sought to outline a 'technology for human affairs', Will Schultz suggested some basic principles which help us – that is anyone – to function fully within our working, as well as our personal, life (Schultz, 1984). These ideas were revolutionary then and still are today. These ideas fit well with the concept of cultures of respect and so we quote them here:

- *Self-regard* – the ultimate basis for my personal and professional success is for me to understand, respect and like myself.
- *Truth* – truth is the great simplifier of personal and interpersonal difficulties.
- *Choice* – I empower myself when I take responsibility for myself.

If these principles are taken as underlying values for helping and supporting those involved in bullying situations, they underpin all actions taken. The helper will be affirming the employee's rights to self-regard and will be working to help this become a reality. A climate will be provided where the employee can become increasingly aware of what is the greater truth of the situation and therefore be more able to become mobilised into action. In turn the whole process will enable the employee to become as free as is possible to make choices and, where willing and supported, to make changes.

The traps

Bullying is an emotive subject. Most people have memories of experiencing terror, helplessness, rage and shame in the face of words or actions levelled at them by others bigger, or more powerful than they. Each person has found his or her own way of dealing with such feelings in order to survive. Sometimes, that includes physical survival but usually the need is for psychological survival with some sense of continuing identity. One survival mechanism is to develop a sense of identity based on some familiar 'roles'. When bullying is in the air, a swift adopting of the stereotypical mindset and behaviour of those roles offers a sense of familiarity – of unselfconsciously knowing what to do and how to be. This can protect against experiencing feelings that threaten to overwhelm and depersonalise.

Figure 11.1 The drama triangle (developed by Stephen B. Karpman, 1968)

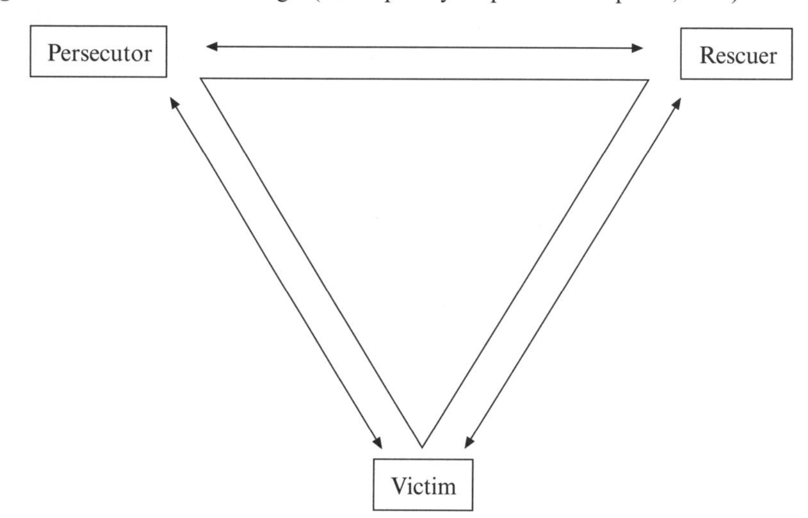

Karpman (1968) identified three roles – victim, rescuer and persecutor – that are almost always taken, usually automatically, in interpersonal situations where power differentials are being exploited. Karpman named his model of role interactions the drama triangle (see Figure 11.1).

Karpman suggested that there are a variety of stereotypical enactments played out, in which people take a familiar role and then switch roles at certain points, thus keeping the drama playing. In this way, some personal and interpersonal truths, which may lie at the heart of the matter, need never be engaged with. These truths are so uncomfortable to some, or all, of the parties that addressing them directly feels unmanageable and perilous.

In a bullying incident, the perceived 'bully' is probably abusing his or her power and persecuting the 'victim'. However, at that point, the bully is also automatically playing the role of persecutor in the way that is well known and favoured because it allows feelings of fear and powerlessness to be held at bay. However, a bully will also know how to take the victim, or rescuer role when the persecutor role no longer achieves its goal.

The victim may or may not be someone who knows how to play the victim role well. Under extreme circumstances and distressing feelings, all of us know how to 'play victim' and to make a feature of our distress when we can no longer hide it. In the bullying 'drama' the object of the bullying is first cast as victim. Then if a supporter is brought into this situation, he or she is cast as rescuer.

The extent to which supporters are familiar with using the rescuer role to bolster their sense of identity – to feel powerful and good about themselves and be protected from feelings of fear, impotence and shame, may unintentionally seal the bullied person into the role of victim. Supporters may even encourage their clients to 'play persecutor' rather than helping them to find their proper power and supporting them to right the situation. It may not be long before the supporter finds him or herself playing persecutor to the bully and feeling victimised by lack of support from the organisation or other members of the team. Subsequently, overwhelmed by resentment, the supporter may feel used by victims who appear to have ignored good advice and services and chosen their own path. The supporter may then become, at least in their heads, persecutory to the victim. Meanwhile the victim may be supportive and protective to the persecutor, thus ending up in the rescuer role.

Case study

A woman had come for counselling after having been severely bullied by a senior manager. The woman was extremely distressed by the bullying and this had had a significant impact on her personal life and working relationships. For several sessions the woman had graphically described her ordeal and, with the support of the counsellor, she had gradually begun to recover. The woman had initially expressed the wish to take out disciplinary proceedings against the manager but she suddenly changed her mind: 'If I make a formal complaint then he may lose his job and I don't want to feel responsible for what would happen to him and to his wife and children.' The counsellor felt let down by the woman – she had listened to all the hurt and anger and now was being told that those feelings were to be put to one side in case the manager's wife and children got hurt. 'You said you wanted him stopped. How is doing nothing going to achieve that?' the counsellor heard herself saying impatiently.

Since none of the protagonists actually enjoy or benefit from being involved in this drama – it is just the way they have learned to be – how can it be substituted by effective action? The beneficial triangle (see Figure 11.2) suggests the factors which underlie the 'roles' described in the drama triangle.

If the experience underlying the roles of victim, rescuer and persecutor are acknowledged and taken seriously, it may be possible to address the painful 'heart of the matter'. Perhaps those things can be said which seem unsayable because of the thoughts and emotions that they could trigger. This depends on at least one of the characters, ideally the supporter but sometimes the victim, having an awareness of self and others which allows a side-stepping of the drama. So how can supporters and helpers avoid the trap of the drama triangle and enable a beneficial triangle to develop?

The supporter has the responsibility of being aware of the potency or the potential for power present in all the players in the situation, where it is being used to persecute, where it is going unnoticed and where it is unsupported or untapped. Supporters also need to be aware of and be relatively at ease with the extent and limits of their own power, personally, organisationally and professionally. In working with employees, supporters need to be alert to their own personal power and to how this is being utilised, ignored or given away to those involved in the bullying. Supporters need to be able to notice how the power that is to be

Figure 11.2 The beneficial triangle

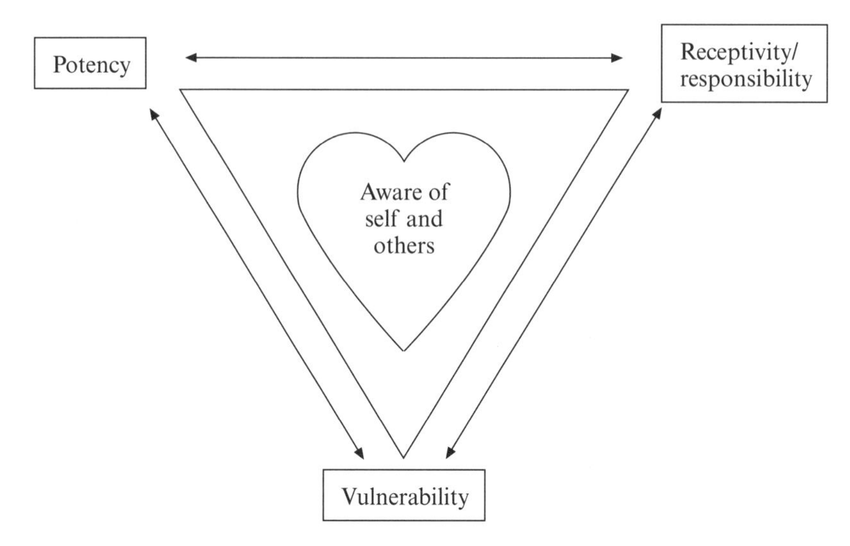

found within formal and informal organisational and personal networks can be tapped in support of the employee's personal growth and development.

What is personal power? Personal power is a description of the individual's ability to make choices and influence others. It is power that each person has, regardless of the formal or informal power and authority conferred by social or organisational roles. Clearly, how people exercise their personal power – appropriately or inappropriately – will be within the context of roles they are given or have taken themselves. A manager will have more opportunity to be empowering or overpowering than will a solitary employee. The employee can be 'kept powerless' with ease by virtue of the role differentiation. However, Schulz's basic assumptions are existential – the individual will always have more power and choice than is easily apparent. With just enough support, challenge and awareness, this can become apparent and the protagonist can become his own agent. To return to the model of the beneficial triangle, the potential for personal power has to be viewed against a background of awareness of the actual and psychological vulnerabilities in the situation: the client's, the helper's, the organisation's and even those of the alleged bully. This requires a vigilant sense of responsiveness and responsibility on the part of the supporter, primarily to the employee, but also to themselves, the organisation and all its members. The dimensions of personal power are illustrated in Table 11.3 which shows how different levels of power are expressed and experienced.

The need for honouring potency, vulnerability and responsibility may require the helper to elicit some facts that do not fit the original story, to challenge the employee's perceived helplessness and to acknowledge the reality of organisational or managerial complicity. This in turn will require helpers to tolerate their own comparative impotence and own feelings of fear and anger. The supporter may

Table 11.3 Dimensions of personal power

Powerless	Powerful	Empowering	Domineering
Ignorant	Informed	Teaching	Instructing
Dependent	Independent	Supporting	Demanding
Vulnerable	Assured	Counselling	Telling
Inadequate	Competent	Nurturing	Punishing
Afraid	Bold	Trusting	Suspicious
Weak	Strong	Co-operative	Competitive
Shy	Confident	Helping	Exacting

even have to risk feeling persecutory towards (and persecuted by) the employee when the employee experiences shame and/or anger at something the helper has hazarded to say.

In tolerating their own discomfort, helpers can act as 'encouragers' to the client to do the same. In words, in attitude or through example, they convey the kind of message that says: 'As a child, you were helpless and at risk – you could not cope with your fear, your shame or your murderous impulses. Your nervous system may be better able to cope now, and objectively, your situation is different and offers you more choices. Together we can look at your problems and with support and understanding you will solve your difficulties.'

Burnout and compassion fatigue

There is a cost to caring. Helpers who listen to stories describing the pain and distress of bullying may themselves feel similar pain and distress. In these situations helpers may find it difficult to maintain their feelings of self-esteem and self-worth when confronted by evidence of the way that human beings can damage and abuse each other. Those helpers who have the greatest ability to feel and express empathy for their clients are the ones most at risk from compassion fatigue. The term 'compassion fatigue' has been used by Charles Figley (1995) to describe the natural consequent behaviours and emotions resulting from knowing about a traumatising event experienced by a significant other – the stress resulting from helping or wanting to help a suffering person. It is perhaps easiest to understand compassion fatigue from the viewpoint of psychodynamic therapy and counter-transference. Counter-transference was once regarded as the therapist's unconscious responses to the client's transference, especially where the transference was connected with the therapist's past experience (Jacobs, 1988). A more contemporary view (Johansen, 1993) regards counter-transference as all the emotional reactions of the therapist towards the client, regardless of their source. Corey (1991) described counter-transference as the process of seeing oneself in the client, of over-identifying with the client, or meeting needs through the client. Through this process of counter-transference the helper can take on the responses of the client as his or her own.

'Burnout', on the other hand, is a state of physical, emotional and mental exhaustion caused by long-term involvement in emotionally demanding situations (Pines and Aronson, 1988). The most widely used burnout scale was developed by Maslach and Jackson (1981). This scale measures three symptoms of burnout: emotional exhaustion, depersonalisation and reduced personal accomplishment. Burnout emerges gradually and is as a result of emotional and physical exhaustion, while compassion fatigue can emerge very rapidly. Figley (1993) designed a self-assessment questionnaire to help assess those employed in dealing with distressed and traumatised clients. This questionnaire differentiates between symptoms that are related to burnout and those that are more suggestive of compassion fatigue.

The role of supervision or consultative support

The role of helping those involved in bullying situations can be very taxing. It renders the helper vulnerable to compassion fatigue and burnout. It requires thought, diplomacy and wisdom for understanding and working in complex systems. It calls for considerable self-awareness, and restraint not to jump into reactive roles. For all these reasons, supporters need their own support system and opportunities to reflect on their practice.

If the helper is a counsellor, psychotherapist or counselling psychologist, he or she will be required, by professional codes of ethics, to be in regular consultative supervision. However, anyone, from whatever working background, who is charged with such work should be provided with consultative support by his or her organisation. This may be offered in a variety of ways but the supervision tasks remain the same.

The Supervision Alliance model (Inskipp and Proctor, 1993) defines consultative (or non-managerial) supervision as:

> A working alliance between a supervisor and a counsellor in which the counsellor can offer an account (or recording) of her work; reflect on it; receive feedback and, where appropriate, guidance. The object of this alliance is to enable the counsellor to gain in ethical competence, confidence and creativity so as to give her best possible service to her client.

The tasks, for both supervisor and helper, which spring from this brief, are:

- *restorative* – the task of affording respite to hard-pressed workers and of using such reflective opportunity for the refreshment of mind and spirit;
- *formative* – the task of reflecting on, and learning from, experience of working with clients and bullying situations and hearing information and getting experience from an experienced colleague;
- *normative* – the task of mutual monitoring that the helper's practice is ethical and within the limits of his or her contract with the organisation/profession.

Inside or outside?

In some ways it is beneficial for supervision to be offered by someone outside the organisation. This frees the helper to reflect fully on the tensions and paradoxes that can arise between the organisation's expectations and the rights and responsibilities of her client/clients. However, it is crucial that the supervisor fully understands the role, rights and responsibilities of all parties, Only then can she provide the helper with non-judgemental reflective space. This work is not like 'proper counselling' or like psychotherapy. It has different aims and objectives and may have a very different contract with the client and organisation. The purpose of supervision is to provide the helper with an opportunity to view her work and her situation 'in the round' and to reflect on what is, and what is not, working well enough. The suggestions, advice or information which she may ask, or which the supervisor may wish to offer, need to be geared to the context.

On the one hand, the supervisor may be someone (other than the helper's line manager) within the organisation. This could be a person more experienced in this work than the helper himself. On the other hand it could be a peer or peers – people engaged in the same work with similar experience to the helper. This has obvious advantages – the supervisor (or supervisors if the work is done in a peer group) knows the context and the tensions involved. Time is saved by not explaining and 'putting the supervisor right' about the context. However, the helper will not have the same freedom to think the unthinkable. There will almost certainly be role tensions – competition, style differences and identifications – which may not be easily spoken about when the parties have other role relationships, day to day.

If supervision is done in a peer group, members of the group need some considerable sophistication, or natural ability, for co-operative work. The tasks of supervision need to be spelt out and group members have to be able to keep to time and task and to be able to support and challenge each other gracefully. Arranging to have annual or biennial reviews, and inviting a consultant supervisor to 'chair', can be a helpful resource.

Perhaps the best solution – time and management permitting – is for helpers to meet in groups for supervision and to have an outside group supervisor to facilitate the group in accordance with a carefully defined working agreement. The group may then grow into a peer supervision group that can meet and work well together, even when there is no outside supervisor.

Supervision as a mirror of the work

Whatever the supervision arrangement, supervisors of those who help in bullying situations have to beware of reproducing the dynamics of the situations being worked with. A well-documented phenomenon in supervision is called the 'parallel or reflective process' (Mattinson, 1975). This process refers to the way that the unrecognised interactions in the helping relationship can be reproduced, without the helper's awareness, in the supervision relationship. So, for instance, the helper who has been bending over backwards to support the victim of bullying may be becoming, unaware to herself, frustrated and judgemental. The unwary supervisor may pick up this frustration from the helper at a non-verbal level. Without being aware of the dynamic, the supervisor may feel equally frustrated with the helper and start to deal with the frustration by overcompensating in a similar manner. At other times, the supervisor might enact his or her frustration by beginning subtly to bully the helper.

In these ways the roles in the drama triangle are reproduced in the new system. This can be useful information for the supervisor and helper if they become aware of it – a parallel process that enlightens the complexities of the work. However, if the supervisor is not aware, then the relationship with the helper will be damaged, perhaps permanently. At the very least, the helper's awareness in triangular situations will not have a chance to improve. Where awareness is missing, skill cannot develop.

The supervision alliance as a model for building a culture of respect

The use of the supervision relationship as a cultural model highlights the need for the supervisor and helper to work together to create a relationship and working alliance in which both can be truthful with each other about any factor which may be affecting good work. This requires the same core qualities from the supervisor (Rogers, 1961) as from the helper:

- The ability to suspend judgement and view the situation, first, through the eyes of the helper.
- The intention of empathic understanding as the major facilitative working medium.
- A willingness to trust and to be trustworthy – for the helper, the working contract and the other stakeholders in the situation whom he may never encounter (manager, client and even the alleged bully). He has a professional responsibility to be advocate for their right to a good service.
- A preparedness to be honest and genuine in offering high quality reflective space, good feedback and professional monitoring.

The supervisor needs to provide these qualities sufficiently to develop and maintain a climate in which the helper can be open and reflective. Through the learning, development and refreshment which this culture enables, the helper will have a model of good helping which will stand her in good stead in her day-to-day work. It is often through the co-operative enterprise of good supervision that new ways of working and fresh understandings emerge. Since the conscious task of building a culture of respect in organisations is so relatively new and developing, this building of new paradigms of practice and understanding is stimulating, feasible and necessary.

The way ahead

Working in an organisation that is involved in the process of building a culture of respect provides an exciting and exhilarating opportunity for counsellors and other supporters. Helpers soon discover that those whose lives have been disrupted by bullying can find relief from a caring professional who can understand and respect their pain, give hope and then get on with the task of motivating and supporting the employee to make the changes that will improve their situation. By understanding the role and accepting that this role carries with it certain responsibilities, the helper is faced with the need to understand his or her own part in fashioning solutions which appropriately recognise the needs and potential of the organisation, the bullied and bullies.

References

Berridge, J., Cooper, C. L. and Highley-Marchington, C. (1997) *Employee Assistance Programmes and Workplace Counselling*, Chichester: Wiley.

Bond, T. (1994) *Standards and Ethics for Counselling*, London: Sage.

Bull, A. (1995) *Counselling Skills and Counselling at Work: A Guide for Purchasers and Providers*, Rugby: British Association for Counselling.

Butler, C. (1999) 'Organisational counselling: the profession's shadow side', *Counselling*, 10, 3, 227–32.

Corey, G. F. (1991) *Theory and Practice Of Counselling Psychotherapy*, Belmont: Brooks Cole.

Edmonds, M. (1991) 'Exploring company welfare', *Employee Counselling Today*, 3, 26–31.

Feltham, C. (1995) 'The Stress of Counselling in Private Practice' in W. Dryden (ed.) *The stresses of counselling in practice*, London: Sage.

Figley, C. R. (1993) 'Compassion stress and the family therapist', *Family Therapy News*, February, 1–8.

Figley, C. R. (1995) *Compassion Fatigue: Coping with Secondary Traumatic Stress Disorder in Those Who Treat the Traumatised*, New York: Brunner/Mazel.

Inskipp, F. and Proctor, B. (1993) *Making the Most of Supervision: A Professional Development Resource for Counsellors, Supervisors and Trainees*, Twickenham: Cascade Publications.

Inskipp, F. and Proctor, B. (1995) *Becoming a Supervisor*, Twickenham: Cascade Publications.

Jacobs, M. (1998) *Psychodynamic Counselling in Action*, London: Sage.

Johansen, K. H. (1993) 'Countertransference and divorce of the therapist' in J. H. Gold and J. C. Neimiah (eds) *Beyond Transference: When the Therapist's Real Life Intrudes*, Washington DC: American Psychiatric Press.

Karpman, S. (1968) 'Fairy tales and script drama analysis', *Transactional Analysis Bulletin*, 7, 39–43.
Lane, D. (1990) 'Counselling psychology in organisations', *The Psychologist*, 12, 540–44.
Maslach, C. and Jackson, S. E. (1981) 'The measurement of experienced burnout', *Journal of Occupational Behaviour*, 2, 2, 99–113.
Mattinson, J. (1975) *The Reflection Process in Casework Supervision*, London: Institute of Marital Studies.
Pines, A. M. and Aronson, E. (1988) *Career Burnout: Causes and Cures*, New York: Free Press.
Rogers, C. (1961) *On Becoming a Person*, Boston, Houghton Mifflin.
Rousseau, D. M. (1995) *Psychological Contracts in Organisations: Understanding Written and Unwritten Agreements*, California: Sage.
Schein, E. H. (1980) *Organisational Psychology*, Englewood Cliffs, Prentice-Hall.
Schultz, W. (1984) *The Truth Option: A Practical Technology For Human Affairs*, Berkeley: Ten Speed Press.
Tehrani, N. (1996) 'The psychology of harassment', *Counselling Psychology Quarterly*, 9, 2, 101–17.

CHAPTER TWELVE

Issues for general practitioners

CHRIS MANNING

Introduction

This chapter considers the importance of bullying in terms of its medical consequences and examines the issues that need to be considered when a general practitioner (GP) is asked to undertake a medical assessment or provide a medical report. The chapter also considers the nature of the GP's role in supporting their patient while they deal with workplace bullying and the benefit of the GP becoming involved in addressing these problems with the organisation. The term 'patient' will be used throughout the chapter to describe an individual who is using the medical system and who is recognised, either to be unwell or to have a formal medical diagnosis. The term does not preclude such terms as 'user' or 'person' which can sometimes be more appropriate where the medical model does not have anything to offer in terms of solving a particular instance of bullying.

Finally, the chapter looks at the range of communication skills that are essential to helping the patient explore their feelings about the bullying and how becoming involved in the process of supporting a bullied patient may become emotionally draining and distressing to the GP.

Clearly, bullying produces measurable physical and physiological effects which, if prolonged, can lead to biochemical and nervous-system changes and consequent diagnosable illness. Although the 'medicalising' of a bullied person's problem can occasionally be helpful when used by a GP to protect the patient, provide recovery time and allow for the mobilisation of resources such as counselling or psychiatric support, this is not always the case. When the 'medicalising' of the problem takes control away from the patient it is neither helpful nor appropriate. The aim of the GP should be to involve the bullied person fully in how they would like the situation to be handled. This requires the GP to have a clear understanding of the patient's right to remain in control but recognises that there may be times when there will be a need to take some of the weight off their patient by giving the appropriate guidance and advice. Some bullied people are able to take more control in their relationship with their GP than others and, since bullying is always disempowering, GPs should be aware that they might not help the situation by taking over too much responsibility from their patient. This is especially true when depression or anxiety are significant factors. While it can be extremely empowering for people to be helped to understand the nature of their situation, the patient may not always have the ability to do so. Distressed patients are not always capable of confronting their bully or organisation and it is important that the GP's own need for action does not place even more demands on an already overloaded nervous system.

Listening to the patient

The traditional strength of general practice has been its emphasis on continuity of care within a therapeutic relationship that lasts over time. Patients get to know their GP and GPs get to know their patients. A clear understanding of the patient's story and version of events is crucial to this relationship; if the GP can listen to someone for long enough, they will find out what is wrong with them. For many GPs listening to the patient 'takes up too much of their valuable time'; although this belief is inaccurate, it illustrates the effects of GPs' stress brought about by working in an increasingly pressurised system. If patients are given the opportunity to express themselves, they are generally capable of making the important points they wish succinctly and with clarity. If GPs take the time actively to listen to patients there are opportunities for them to make accurate assessments of their patients' psychological functioning through a careful observation of non-verbal communication, including body language, voice tone, pitch and volume (Egan, 1990).

The approach adopted by the GP when talking to the patient is critical in the communication process and the best results are achieved when there is good eye contact and open questions are used. Closed questions that simply produce 'yes' or 'no' responses lead to a very narrow investigation of the problem. The use of open questions and paraphrases, such as 'so what you are saying is that you have been bullied by your manager on several occasions and that it is particularly difficult in team meetings?' is more likely to enable the patient to describe the complete problem than being asked 'did your manager upset you?' Through careful listening the GP can begin to understand the patient's core values and beliefs by looking for the emphasis placed on certain words or phrases. In the following examples the words in italics were emphasised by the patient: 'That is *not the right way* to treat people', 'I know *it's not just me* that is affected' and 'My boss is a complete *bastard*'. Reflecting back, the emphasised words or phrases can help the patients to explore the most important aspects of their experience without the need for the GP to take control over the direction of the enquiry (Elgin, 1995).

Listening and responding to the emotional content of communication, as well as to the narrative description of the events, is important if an understanding of the full extent of the impact of the bullying is to be achieved. In the GP's busy day it is essential that the patient is helped to communicate their most important concerns at the beginning of the consultation otherwise the 'while I am here doctor' statement will be uttered just as the patient is about to leave the consultation room. Helping the patient to identify the important issues for consultation at the outset supports the therapeutic relationship, improves patient confidence and makes the best use of the time available.

Unfortunately, there are many things that can get in the way of good communication. Two of the more serious problems that are particularly relevant when dealing with bullying are when the GP is suffering from burnout (Maslach and Jackson, 1981) or compassion fatigue (Figley, 1995). When a GP is suffering from burnout, normally brought about through overwork, the three classic symptoms of emotional exhaustion, depersonalisation and reduced personal accomplishment can prevent the GP from meeting the needs of the patient. Compassion fatigue is most likely to affect medical professionals who develop a strong empathetic understanding of their patients with the result that the GP experiences similar feelings of helplessness, confusion and distress as those being experienced by their patient. GPs who are suffering from burnout or compassion fatigue are unlikely to be able to maintain the appropriate levels of involvement and professional objectivity which are essential when dealing with bullying. It is ironic that many of those with the most ability in a caring capacity should themselves be most prone to suffer personal distress as a result.

Taking time to listen to a patient is rewarded in the following case which occurred in an NHS practice. The case illustrates how the patient's full story was only teased out following the involvement of a number of people in what was a haphazard rather than managed process. The story demonstrates the importance of everyone involved in supporting the patient being able to communicate with each other.

Case study 1

Monica came to the surgery with severe headaches. She had been involved in a car accident nine months previously and had sustained a whiplash injury that required a month off work. Her outpatient appointments and subsequent investigations meant that she needed more time off work, as did the physiotherapy and reviews by her GP. As time progressed it became increasingly obvious that Monica was becoming depressed and needed to talk about the consequences of her accident in terms of her work rather than of her physical symptoms.

One day Monica attended the clinic for a routine contraceptive pill check with the practice nurse who found her blood pressure raised; Monica was told she must see the doctor before the tablets could be prescribed. When the receptionist told her that there was a two-week wait for an appointment, she burst out crying and was taken to see another GP who reassured her that her blood pressure was normal and told her to come back and see her usual GP in a fortnight. When Monica returned to the reception to book her appointment, she was still distressed and the receptionist spoke to Monica's own GP and asked him to see her as an emergency at the end of the surgery. By the time Monica saw her usual GP, she was much calmer. Monica's GP noted the raised blood pressure and suggested that it might be caused by the stress she had been experiencing. Monica said that she thought that she was 'going mad'.

She had always thought of herself as a stable person, and had never previously doubted her own competence. She then became very tearful again and said that her line manager had described her as 'feeble' and that she should 'pull herself together'. Monica was told that she had exceeded her limit in terms of sickness absence and that the disciplinary procedure would be applied unless there was 'substantial improvement' in her work record and 'attitude'.

In Monica's case a number of people clearly demonstrated that they were sensitive to her needs. However, this is not always the case – some GP surgeries have their own bullies and the traditional hierarchy or 'pecking order' of professional and support staff can prevent patients from receiving the respect and care they deserve.

Bullying as a cause of ill health

All forms of bullying, whether personal or corporate, can cause stress. Stress, in the short term, raises the levels of hormones of the adrenal medulla, adrenaline and nor adrenaline (epinephrine and nor epinephrine) with consequent rises in levels of pulse rate, blood pressure and alertness (Arroba and James, 1987). Traditionally adrenaline and nor adrenaline have been regarded as the 'fight or flight' hormones (Cannon, 1927). The description should have also included 'freeze', another acute stress-coping mechanism which has been observed in many people faced with a seemingly insurmountable problem (Raphael, 1986). In the longer term, stress produces an increase in secretion of the adrenal cortical hormone cortisol; this, in turn, is linked to an increased incidence of coronary and peripheral arterial disease (heart attacks and strokes), impaired immunity, anxiety-depression and osteoporosis (Dinan, 1999).

Case study 2

A woman came into the surgery; she had an extremely important job in the City of London and had always seemed confident and capable. Immediately she walked into the consulting room it was clear that there was something wrong. She said that she had been feeling run down and kept on having colds. A skin condition that had not troubled her for several years had broken out again and she said that she was unable to sleep. When asked if there was anything troubling her, she immediately burst into tears. It seemed that she had taken on a new job but that her manager was not providing her with an appropriate level of support or guidance. She was continually being criticised in public for not producing what he required, despite the fact he never actually told her what he wanted. She had become very tense and found herself getting angry and irritable with family and colleagues. She was constantly tired and lost over a stone in weight. Her self-confidence was low and she found that she was unable to make decisions because she was afraid of making a mistake. In a period of three months this previously confident and capable woman had become physically and psychologically ill owing to the continual bullying by her manager.

Unfortunately, many doctors dealing with increasing demands from the NHS come to believe that they do not have the time to deal with social problems such as bullying. Looking at scarce resources, including their time, it is important for GPs to reflect upon the fact that between 30–40 per cent of all sickness absence from work is due to some form of mental or emotional stress (O'Leary, 1993). Taking the time to address issues that are likely to have a direct impact on mental health is a more positive approach than merely to address the physical symptoms. It is interesting to note that hospital investigations of patients with stress-related symptoms are not very effective in identifying causal factors, with 50–60 per cent of cases remaining undiagnosed (Wessely, 2000). This medical approach is wasteful of scarce hospital beds when the problem may more readily be established by simply talking to the patient.

The effect of litigation on the GP

As more employees are prepared to consider taking legal actions against their employer for bullying it is important for GPs to be aware of the effect the legal system can have on the well-being of their patients. Studies of victims of stress and trauma have shown that involvement in a litigation case can be extremely stressful, particularly when the case takes a long time to come to court (Pitman et al., 1996). A GP needs to be able to help the patient deal not only with the bullying at work but also with the problems of dealing with the legal system which may take years to bring a case to court. The range and variety of litigation cases that have been reported in the past few years can help the GP to recognise the types of workplace bullying which may occur.

A case involving the North Yorkshire constabulary showed that a culture that encouraged bullying had been allowed to develop in the Harrogate police station in which policewomen and new recruits were systematically bullied and humiliated. In the investigations that took place in 1996, it was shown that new recruits were forced to run around the police station with bulldog clips fastened to their nipples and one officer was kept in a dog kennel for three hours for wearing the wrong shirt (*Yorkshire Post*, 1996).

While the case described above may appear extreme, it is by no means unusual (Field, 1996; Randall, 1997). A TUC report estimated that five million employees are bullied at work annually (TUC, 1998). However few of these employees actually report the bullying to their employers. A survey of 157 employers (IRS, 1999) showed that, while the number of formal bullying complaints was low, most organisations recognised that employees were reluctant to report bullying, especially where there are no mechanisms for them to do so. The high incidence of bullying in the workplace and the lack of appropriate organisational responses results in employees becoming physically and psychologically distressed. For many employees their GP is their only source of support in dealing with a difficult problem.

Bullying and its relationship to anxiety and depression

An Institute of Personnel and Development survey conducted in 1996 found that one person in eight had been bullied at work within the last five years. Such harassment 'extracts a high price from both employees and employers alike'. If bullying at work is associated with job dissatisfaction, absence, poor performance and turnover, the distressed employee will attend the GP's surgery with symptoms of stress, depression and anxiety, together with their related physical and occupational manifestations (see Table 12.1).

In a survey of staff in one NHS community trust, Quine (1999) found that over a third (38 per cent) reported being subjected to one or more forms of bullying in the previous year and 42 per cent had witnessed the bullying of

Table 12.1 The impact of bullying

Symptoms and signs of depression and anxiety	
Low mood	Tiredness and fatigue
Loss of pleasure in life	Tearfulness
Nightmares	Irritability
Suicidal thoughts	Anxiety and agitation
Poor concentration and memory	Panic attacks
Sleeplessness	
Physical symptoms	
Frequent illness	Loss or gain in weight
Colds	Bowel upsets
Cystitis	Palpitations
Thrush	Skin problems
Cold sores	Eczema
Appetite change	Psoriasis
Decreased libido	Aches and pains
Ejaculatory problems	Ulcers
Behavioural consequences at work	
Absenteeism	Resignation
Shattered self-confidence	Poor performance
Low commitment	Lack of imagination
Accidents	Increase in errors and mistakes

others. Sixty-seven per cent had tried to take action about the bullying when it occurred but the majority (74 per cent) was not satisfied with the outcome. Quine provides the following explanations for the association of bullying with anxiety and depression:

- Being bullied leads to psychological ill health and decreased job satisfaction.
- Some people may be more likely to report being bullied than others; these may be people with a pessimistic outlook.
- Being depressed, stressed or anxious may cause a person to be bullied by people who choose 'weaker' people as their victims.
- Anxiety and depression weaken a person's ability to cope with stressors such as bullying or make them more likely to perceive other people's behaviour as over-hostile or critical of them.

Some professions appear to be particularly vulnerable to bullying (Cooper and Hoel, 2000). Over the past few years there has been a large increase in the numbers of teachers seeking medical advice for stress and many of them request support for applications for early retirement on medical grounds, reporting that their stress was being caused by bullying. Organisations have their own history and style of operation and these organisational cultures can determine the nature of the bullying that takes place. In a study of organisational cultures, Wright and Smye (1998) described three types of corporate culture and the style of abuse which were adopted:

- The *win/lose culture* which forces people to strive against their colleagues rather than with them.
- The *blaming culture* which makes people afraid to step out of line.
- The *sacrifice culture* which involves people putting work above their social and personal lives.

A GP caring for a patient working in a destructive culture is faced with the dilemma of merely dealing with the symptoms of the problem or, with the agreement of the patient, tackling the underlying problem with the organisation. Sometimes the real problem is not easy to identify, particularly when the patient has a reason to hide their true fears, as was the case in the following study.

Case study 3

Jack was an experienced and well-liked line manager with 25 years of experience in the same company. Over a period of a year, he became increasingly withdrawn and aloof. He would walk past colleagues and friends in the corridor and grunt a greeting, rather than stop and engage in his usual friendly conversation. It was also noticed that he was losing skills for which he was widely respected: his ability to engage those with whom he worked and his high level of personal productivity. Concerns were expressed at his performance and he was requested to attend a 'friendly' hearing. At this stage, he visited his GP who thought he was depressed. He was reluctant to take medication and said that he would be all right now that he had mentioned the problems at work but was adamant that he did not want any medical assistance or documentation. Problems at work continued until, some six months later, he was summoned to a disciplinary hearing for his unreasonable behaviour towards others who resented his indifference to their expressed concerns and his increased aggressiveness. After a very prickly session, in which Jack was very defensive, one member of the committee noticed that he was fumbling to find the glasses that he had placed on the table in front of him during the interview. When asked about this and whether he was over-stressed, Jack broke down completely and through his tears admitted that he was losing his sight. He was then referred back urgently to his GP.

It seems hard to believe that Jack's friends, manager, family and GP missed such an obvious problem yet, because of Jack's denial and an unwillingness of others to deal with issues for fear of infringing Jack's rights, it was possible for the problem to be missed and for his behaviour to be misinterpreted. When an employee needs to keep their job, there is a huge pressure to hide anything which might get in the way. A working environment in which employees are constantly under 'threat' of job insecurity and constant change, inconsistent personal managerial decisions can result in behaviours which deny or hide personal problems and which may themselves increase the likelihood of the employee being made unemployed. GPs need to be aware of these fears when listening to their patients and then to look for a solution which addresses the underlying fear, as well as the medical condition.

Physician heal thyself

Nurses, doctors and other care professionals are not immune to bullying. The need for increased 'performance' with a very low level of investment compared to other western European countries, has affected the way in which many trusts and health authorities have treated their employees. A study of health professionals has shown that when the effects of personal vulnerability to psychiatric disorder and ongoing personal stress are controlled, stress at work has the greatest contribution to anxiety and depressive disorders in these workers (Weinberg and Creed, 2000). In another survey of 800 GPs Griffiths (2000) said that 200 were suffering from depression, a further 400 said that they were afraid that they would suffer from a stress-related illness within the next five years and 40 admitting to abusing alcohol.

As the largest employer in Europe, with 750,000 employees, the NHS is remarkable in that it has no nationally co-ordinated employee assistance programme or external occupational health service for those who work within it. At a conference at the Royal College of GPs in 1996 and then again at the Royal Society of

Medicine in 1999, there was a strong appeal for such services to be set up. The personal experiences of doctors recounted within the safe and anonymous surroundings of self-help groups, such as the Doctors' Support Network, give an indication of how many of these professionally-trained and medically-qualified healthcare workers attribute their mental illness to the stresses and bullying encountered in their training and clinical careers. The lack of co-ordinated and confidential support from the NHS as an employer leads to the intensification of these problems. This bullying described by the professional contacting the network includes:

- The increasing responsibility of the job with a decreasing authority to act.
- The state of constant change with little involvement of those most affected.
- The introduction of 'gagging' and 'whistle-blowing' contract clauses.
- The denial of ongoing professional development.

Where doctors are embittered by their own situation there is an increased possibility for them to project their frustrations on to their patients and colleagues. It is not uncommon to find GP practices where there are incidences of harassment, verbal aggression, unreasonable workloads, exclusion from the decision-making process or unwarranted undermining of colleagues' experience or ability. Letters received at Depression Alliance – the largest self-help organisation in Britain for those with depression – indicate how attitudes acquired by GPs during their training, or resulting from the slower process of burnout, can affect their judgement. Such statements as 'pull yourself together,' 'you've just got to face the fact that you are not going to be one of life's little sunbeams' and 'I think that what you have to say is completely irrelevant' do not exactly move the issues forward in a positive manner. Where a patient is faced with a GP who responds to real concerns in this way and where there is a feeling that the abuse which is being experienced in the workplace is also occurring in the GP's surgery, the patient should take this up with the GP. Challenging a GP is extremely difficult for people to do, particularly those who are already feeling disempowered by bullying. For many people the GP is a revered figure who is always right. If a patient has the courage to challenge a GP, it is important that the GP takes the time to consider whether there is any truth in the complaint and, where appropriate, tries to be more responsive to the needs of patients.

The GP as advocate: 'perception is reality'

A GP can become an important advocate for a bullied patient. However, it is important not to precipitate or become involved in advocacy without establishing the facts. It may be obvious to the GP when a patient has an 'axe to grind' but it may be more difficult to identify those people who are so distressed that their judgement has become impaired. In the following case a GP takes the side of his patient, unaware that there is another side to the story.

Case study 4

The patient was a manager of a large group of people. He had always done well in the past and the sales figures for his section were among the highest in the company. He was then moved to a different location that had been doing less well. He was given a deputy manager who had been brought in from another organisation. The manager had a tendency to lead from the front and had a strong need to be in total control. When he was unable to turn the sales around he began to put pressure on his staff and on the deputy manager in particular. Some of the techniques he used could have been described as bullying. The deputy manager went to the personnel manager and complained but, instead of dealing with that situation, the personnel manager agreed with the deputy manager and put in her own complaint against the manager. The manager was called to a meeting with his boss at which the catalogue of complaints of bullying was presented. The manager then went off work with stress.

The GP wrote to the company, not having fully explored the precipitating events. All the GP knew was that the manager had been summoned to a meeting where he was accused of not managing his staff and that he was not given the opportunity to respond. The GP's perception was that the manager was being bullied and that he was experiencing stress as a result. The manager was off work for over six months, during which time attempts were being made to rehabilitate him, however the GP's view that the organisation was totally to blame for the situation made it difficult for the manager to accept his part in the problem.

The GP and patient can, by working together, help to re-establish self-esteem and self-worth through the nurturing of psychological well-being. For a GP this restorative process can be both exciting and rewarding as it is only when the patient has regained their mental health that they can, with support, challenge the bully and their behaviours. Such re-empowered individuals are grateful and the perfect antidote to those patients who GPs may label as 'heartsinks'. It is time for GPs to acquire new psychological skills and ways of working and the issue of bullying is fertile ground for those who wish to break free from the traditional 'doctor knows best' approach.

Case study 5

A GP had been seeing one of his patients who had described a series of bullying episodes that had involved her line manager. When he noticed that Louise was no longer wearing her wedding ring, he commented upon this, whereupon she admitted that she had gone off sex; her husband, she was sure, was seeing someone else. On further questioning, she admitted to having had suicidal thoughts and nightmares. The GP asked Louise to fill in some questionnaires. Louise was adamant that she would never actually commit suicide but recognised that she needed help. The GP agreed to see her in a few days and booked her in to see the practice counsellor. Louise phoned the next day to thank him and to say she was feeling much better. The GP asked her to start keeping a daily diary of her thoughts and to compose a letter that she would write to her employers if she were the doctor.

Keeping a diary can be empowering for patients who have been bullied. The diary allows them to tell their story and the process of describing what has happened during the bullying, and then having the story reflected back from another's point of view, can be liberating. For the GP, the approach has the added advantage that the work involved in taking the history and its clarification is minimised. The subsequent document can form an excellent basis for moving the issues forward very quickly and this mutual involvement can be both satisfying and empowering. Instead of the GP using multiple consultations that may not deal with the underlying cause of the medical problems, the real problems are identified and progress is made.

In consultation techniques such as as stress-mapping, problem-solving and solution-focusing interventions are useful in assessing the locus-of-control issues and to establish where the individual's coping skills lie. These interventions have the added benefit that they stimulate ideas that are novel. Too often GP consultations have started with 'what do you think I should do doctor?' and ended with the doctor saying 'I think you should . . .' 'Could' is a far better term to use as it is less prescriptive and more open-ended. It can also be useful to involve the patient's family or carer in consultations and, if the patient is willing, to audio-tape the consultations when the issues are being teased out.

It can be very helpful to ask people to fill in self-report questionnaires such as the HAD (hospital anxiety and depression rating scale) or the Beck Inventory to enable a more detailed assessment of the depressed mood to be achieved. The SASS (self-assessment social functioning scale) questionnaire can be used to measure social functioning. All these questionnaires, together with physical examination if indicated, help to establish the current state of mind and distress. It is helpful to complete these when bullying is first indicated since the scores can serve as a benchmark for monitoring the outcomes of any interventions and may have a medico-legal importance in the future.

When contacting an organisation, a GP should be satisfied with the quality of the information that they are going to use in their report. When composing a medical letter to be sent to an organisation, the GP should ask the patient to keep a diary providing information on the inappropriate behaviour. Where possible the diary should record events over several weeks and should include:

- a record of incidents including times, dates, details;
- any inappropriate contacts out of working hours;
- any responses that they took at the time of the incident;
- the names of witnesses to the bullying incidents;
- copies of any appraisals or correspondence;

- their job description and current responsibilities;
- their company policy on bullying and harassment.

The diary provides tangible evidence of the nature of the bullying which can be augmented by the results from the questionnaires. In order to protect the patient from further abuse, the following advice can be given:

- Avoid situations where they are alone with the bully.
- Refute the bully's claims in writing.

After this process most victims have a far better understanding of their predicament and an increased feeling of being in control. This sometimes enables the patient to address the bullying themselves by taking their evidence to their trade union, human resources or occupational health department without the direct involvement of the GP.

Confronting the bullying

Bullies and organisations frequently fail to recognise, or take responsibility for, the health and well-being of the victims of bullying. It is important therefore that the GP recognises their duty of care to their patient by taking the opportunity to make the bully and organisation aware of the damage that is being inflicted on their victim. Through this communication the GP makes sure that the legal responsibility for taking action to stop the inappropriate behaviour is fully recognised by the organisation. Without powerful feedback on the inappropriateness and damaging nature of bullying, the perpetrators of bullying and the organisations where they work will be under no pressure to change. A bully left unchecked will continue with their antisocial and harmful behaviours leading to the likelihood of causing distress to an increasing number of people. In general, it is not a good idea for a GP to confront the bully directly. Bullies may have personality problems where any challenge can result in the victim being punished even more for 'telling lies to the doctor'. A better strategy for the GP is to ask the patient whether this is happening to anyone else at work and to make bullying a line management issue. If the bully is a line manager, then it is necessary to identify a superior to that person. It is rare to have a situation where the patient has no one in an organisation that they can trust or respect and in the case of larger organisations a union representative may be the most appropriate person.

Caring for the bully

It may seem strange that the GP should also become involved in helping and supporting bullies. In a lecture given in 1994, the American behavioural change therapist Ernie Larson said 'what we live with we learn, what we learn we practise, what we practise we become, what we become has consequences'. This is another way of saying that practising a technique builds up neuronal networks and associated behaviours that become increasingly hard to break. Larson went on to say that it is not practice that makes perfect but only perfect practice that makes perfect.

An Australian study of schoolchildren (Ferero, 1999) found that bullied students who also bullied others were at the significant psychological risk of suffering anxiety, depression or committing suicide. A study from Finland (Kaltiala-Heino et al., 1999) corroborated this study and, further, suggested that psychiatric intervention should be considered for the bullies as well as their victims. When researchers looked at depressive symptoms, it was found that students who had bullied were most likely to consider committing suicide. One possible explanation is that happy people are unlikely to want to make others unhappy through bullying. There is always the risk that, in undefended and lonely moments, bullies may recognise what they have done and feel deep remorse. It may be transient, but it can be very deep. Within my own practice I know of the suicides of two men of this type. The cases were more difficult to handle because each man had claimed the cause of their behaviour to be severe depression, yet they did not wish to receive any help in treating the depression. Unfortunately, this presents a dilemma for a GP since, to quote the old adage 'the light bulb must want to be changed' and producing such change may mean behaving very differently. In this context, bullying, or any other established behaviour for that matter, should be seen as no different to alcoholism with remedies as robust and supportive as those employed in the Alcoholics Anonymous 12-steps programme. A bully may be resistant to any suggestion of change and afraid of giving up the unacceptable use of power that previously provided recognition and reward. In Britain, psychological interventions are not readily available on the NHS and people may be unwilling to embark on courses of action for which they might have to pay. The cost, however, becomes less of an issue when the bully faces the threat of dismissal if they continue 'offending'.

Discussion

In conclusion, this chapter has considered the importance of the stress caused by bullying at work and the economic folly of ignoring the scope of this problem. It has suggested that the time has come for organisations and employers to recognise the GP's duty of care to patients. In order to meet that duty issues within the workplace may need to be recognised and addressed. The employer also has a duty to provide safe working conditions which require employers to take reasonable steps to tackle anything that adversely affects the physical or mental health of the workforce. Whatever the victim feels, bullying is still bullying. While framing the person in a medical model, it is not desirable to label and treat someone as ill when they are simply registering the toxic influence of another, whose behaviour must be made answerable to change both by themselves and those who have influence over them.

The NHS as an organisation has experience of bullying by some of its employees and users. The consequences of this culture may result in the unfair treatment of people who are already victims of workplace bullying. These internal issues must be acted on robustly if Britain wishes to move towards a position in which it is regarded as first class in terms of mental health.

Where to go for help

In addition to referral to counsellors, clinical psychologists, community psychiatric nurses and psychiatrists, GPs can contact other agencies. It is worth considering producing a list of agencies to be handed out as required.

The following organisations are helpful:

- Social skills training courses at local colleges – here patients can learn skills including assertiveness, communication and self-awareness.
- Depression Alliance has a country-wide network of self-help groups, informational and support literature (tel: 020 7633 0557).
- Citizen's Advice Bureaux provide helpful support and advice for professionals and patients when preparing reports.
- ACAS provides arbitration and conciliation advice and booklets online and at local offices around the country: www. acas.org.uk
- UK National Workplace Bullying Advice Line (tel: 01235 212286).
- The Andrea Adams Trust provides a helpline for advice on bullying (helpline tel: 01273 704900, Monday–Thursday, 10 am–4 pm).
- The Industrial Society exists to improve working life: www. indsoc.co.uk
- TUC 'Know Your Rights' Line (tel: 0870 600 4882).

For GPs wishing to learn new skills in dealing with those with mental health problems: www.primhe.org provides information and links to other helpful sites and courses.

References

Arroba, T. and James, K. (1987) *Pressure at Work: A Survival Guide*, London: McGraw-Hill.

Cannon, W. B. (1927) 'The James Lange theory of emotions: a critical examination and alternative theory', *American Journal of Psychology*, 39, 106–24, 333.

Cooper, C. L. and Hoel, H. (2000) 'Destructive interpersonal conflict and bullying at work', press release, 14 February 2000.

Dinan, T. (1999) 'The physical consequences of depressive illness', *British Medical Journal*, 318, 826.

Egan, G. (1990) *The Skilled Helper: A Systematic Approach to Effective Helping*, Pacific Grove, CA: Brooks Cole.

Elgin, S. H. (1995) *You Can't Say That to Me!: Stopping the Pain of Verbal Abuse*, New York: Wiley.

Field, T. (1996) *Bully in Sight: How to Predict, Resist, Challenge and Combat Workplace Bullying*, Wantage: Wessex Press.

Figley, C. (1995) *Compassion Fatigue: Coping with Secondary Traumatic Stress Disorder in Those Who Treat the Traumatised*, New York: Brunner/Mazel.

Griffiths, H. (2000) GP survey, *Pulse*, February.

IRS (1999) 'Bullying at work: a survey of 157 employers', *Employee Health Bulletin*, 8, 677, 4–20.

Kaltiala-Heino, R., Rimpelä, M., Marttunen, M., Rimpelä, A. and Rantanen, P. (1999) 'Bullying, depression, and suicidal ideation in Finnish adolescents: school survey', *British Medical Journal*, 319, 348–51.

Maslach, C. and Jackson, S. E. (1981) 'The measurement of experienced burnout', *Journal of Occupational Behaviour*, 2, 2, 99–113.

O'Leary, L. (1993) 'Mental health at work', *Occupational Health Review*, September/October.

Pitman, R. K., Sparr, L. F., Saunders, L. S. and McFarlane, A. C. (1996) 'Legal issues in post traumatic stress disorder' in B. van der Kolk, A. C. McFarlane and L. Weisaeth (eds) *Traumatic Stress: The Effects of Overwhelming Experiences on Mind, Body and Society*, New York: Guidford.

Quine, L. (1999) 'Workplace bullying in a NHS community trust: staff questionnaire survey', *British Medical Journal*, 318, 228–32.

Randall, P. (1997) *Adult Bullying: Perpetrators and Victims*, London: Routledge.

Raphael, B. (1986) *When Disaster Strikes: A Handbook for the Caring Professional*, London: Unwin.

TUC (1998) 'Beat bullying at work' campaign, London: TUC.

Weinberg, A. and Creed, F. (2000) 'Stress and psychiatric disorder in healthcare professionals and hospital staff', *Lancet*; 355, 533–37.

Wessely, S. (2000) personal communication.

Wright, L. and Smye, M. (1998) *Corporate Abuse*, New York: Simon and Schuster.

Yorkshire Post (1996) £150,000 pay-out to stress case policeman, 18 September.

CHAPTER THIRTEEN

Bullying is a trade union issue

TOM MELLISH

Introduction

This chapter looks at the important role the unions play in reducing the occurrence and impact of bullying at work. The TUC and its affiliated unions' work on stress identified the fact that bullying was one of the major causes of stress at work. In the words of the campaigning journalist Andrea Adams (1992), workplace bullying is 'offensive behaviour through vindictive, cruel, malicious or humiliating attempts to undermine an individual or group of employees'. Andrea Adams went on to describe workplace bullying as:

> persistently negative attacks on personal and professional performance, typically unpredictable, irrational and often unfair. This abuse of power or position can cause such chronic stress and anxiety that the employees gradually lose belief in themselves, suffering physical ill-health and mental distress as a result.

These definitions have been widely accepted throughout the trade union movement. The TUC's view that workplace bullying is a health and safety issue is fully explained in the chapter, as is the need to develop effective organisational policies and procedures aimed at reducing the likelihood of bullying occurring. Where bullying has occurred, the approaches that are required to reduce any further distress are described, as is the role of the union representatives in supporting the employee throughout all the stages of investigation and possible legal action.

Stress and bullying

Stress is the term used to describe distress, fatigue and the feeling of being unable to cope. When the employee's available resources do not match the demands and pressures placed on them, stress can occur, endangering the employee's

Table 13.1 The main hazards of concern to workers (survey of safety representatives, TUC, 1998)

Hazard	% cited by safety reps
Overwork or stress	77
Display screen equipment (DSE)	48
Slips, trips, falls	46
Back strains	44
Repetitive strain injuries (RSI)	37
Chemicals or solvents	46
Noise	30
Violence	28
Working alone	28
High temperatures	27
Long hours of work	25
Machinery hazards	24

Note: percentages do not total 100 per cent because reps could tick up to five main hazards

health and well-being. In the short term stress is debilitating, but in the longer term it can kill.

Stress at work affects the mental and physical health of over half a million British workers every year. This has economic implications for businesses and for the nation as a whole. The 1995 self-reported work-related illness survey (Jones et al., 1998) suggests that a quarter of a million workers suffer from occupational stress and the same number have an illness caused by stress at work. Between five and six million working days are lost every year owing to workplace stress and its effects. The Department of Health (Jenkins and Warman, 1993) estimates that the cost of sickness absence for stress and mental disorders in Britain is more than £5 billion a year.

Bullying, according to both the TUC (Tudor, 1996) and surveys of trade union safety representatives (Kirby, 1998) was one of the major causes of stress. In each of these surveys (see Tables 13.1 and 13.2) safety representatives were asked to assess the major health and safety issues for themselves and their members. Over 7,000 safety representatives responded to each of these surveys. In each survey stress was identified as the priority issue, increasing from 61 per cent in 1996 to 77 per cent in 1998.

Twenty-one per cent of safety representatives identified bullying as a problem in their workplace and recognised that bullying was a major cause of stress. These results indicate a significant increase in bullying since the 1996 TUC survey in which bullying was only identified as a concern by 14 per cent of respondents. The problem of bullying in the workplace therefore appears to be getting worse. This finding was further reinforced by the TUC's 'Bad bosses hotline' week in December 1997. The TUC ran the telephone hotline looking for evidence of low pay, long hours, health and safety disaster areas and workplace bullying. During the five days the TUC received nearly 2,000 calls from workers complaining about being bullied. This represented 38 per cent of the total of calls received that week.

Table 13.2 The five main hazards of concern to workers (by sector)

Sector	Main concern	2nd concern	3rd concern
Health services	Overwork or stress 82%	Back strains 74%	Violence 44%
Distribution, hotels, restaurants	Overwork or stress 77%	Back strains 72%	Slips, trips and falls 72%
Banking, finance, insurance	Overwork or stress 84%	DSE 84%	RSI 71%
Voluntary sector	Overwork or stress 90%	DSE 60%	Violence 46%
Education	Overwork or stress 88%	DSE 44%	Slips, trips and falls 38%
Energy and water	Slips, trips and falls 73%	Overwork or stress 72%	DSE 49%
Central govt	Overwork or stress 90%	DSE 83%	RSI 64%
Local govt	Overwork or stress 81%	DSE 61%	Slips, trips and falls 45%
Transport and communications	Overwork or stress 78%	Slips, trips and falls 63%	Long hours of work 56%

Note: percentages do not total 100 per cent because reps could tick up to five main hazards.

The TUC hotline showed that bullying took place in two main ways:

- *Individual bullying*, sometimes taking place in well-managed companies, where individual managers get away with bullying their staff and making their lives a misery. Some callers worked in firms that operate dignity at work policies but where the behaviour of some managers simply slipped through the net.
- More commonly, callers telephoned about *corporate bullying* where companies had rough treatment built into their business plans and middle and junior managers felt they had no other choice but to treat their staff harshly.

An example of the latter is given by 'Sanjay' who works as the manager in a well known leisure shop:

> My staff are expected to work a nine-hour day with only a half-hour unpaid lunch-break for just £16. This is despite larger than expected profits. You've got to be Mr Nasty if you want to get on – that's the word from the top.

In October 1998, during a week of TUC activity as part of the TUC's 'No excuse – beat bullying at work' campaign, over 1,300 calls were received by the TUC's Bullying Information Line. In addition there were a similar number of personal visits to the TUC information booths that had been set up in a number of city centres around the country. Callers to the information line ordered leaflets giving them information on how to tackle workplace bullying and where to get help.

The following three examples of bullying are taken from the TUC guide, *Tackling Bullying at Work*.

Case study 1

'Sarah', store supervisor in a high street clothes shop:

> He spent most of his time playing practical jokes on the younger women staff – even though none of them saw the funny side. He would shout out their dress size across the shop and when people left the job he would ruin their leaving parties by threatening to play practical jokes on them. He ruined mine by dragging me across the shop floor and pushing my face into a sink of cold water.

Case study 2

Raymond, group editor for a newspaper company:

> My ex-boss used to manage his staff by humiliation. He would make people, who did not reach the impossible targets he set, stand in the corner wearing a dunce's hat. The worst thing was he was convinced his behaviour made him a good boss – that it would increase productivity. But the staff were terrified – some of them literally jumped every time he walked in the room.

Case study 3

Paul, trainee sales representative for a publications company:

> As an ex-police officer I have been through some pretty rough times, but nothing compares to the treatment dished out to me and my colleagues by this man. He would time me when I went to the toilet and, when we met up on field sales days he would take each of us outside and tear strips off us. One day, when I hadn't made the number of sales that I was supposed to, he actually threatened to assault me. But he was very clever, he rarely threatened us in front of other staff and if you tried to approach the subject with supervisors they would deny bullying was even a possibility.

The extent of the problem

The cost of workplace stress to the British economy and to the individuals who suffer from it is enormous and growing. Employers who fail to tackle bullying pay a price. Their cost is in lost time because staff are affected by stress and ill-health; lost incentive because morale is low; reduced work output and quality of service and lost resources because people who are trained and experienced leave the organisation. A report from the Institute of Management (1996) said an estimated 270,000 people take time off work every day because of work-related stress. This represents a cumulative cost in terms of sick pay, lost production and NHS charges of around £7 billion annually.

Cary Cooper, from the University of Manchester Institute of Science and Technology (UMIST), says that bullying is a very significant factor in stress at work that could account for between a third and half of all employment-related stress (*Guardian*, 1 December 1998). He has estimated that some 40 million working days a year are lost because of bullying at work, at a cost to the economy of £1.3 billion a year.

UNISON, the public service union, defines workplace bullying as 'offensive, intimidating, malicious, insulting or humiliating behaviour, abuse of power or authority which attempts to undermine an individual or group of employees and which may cause them to suffer stress' (UNISON, internal paper). UNISON goes on to show that bullying is a problem that has increased in recent years. There are indications that cost-cutting by employers, in terms of both staff and resources, along with factors such as the trend for pay to be linked to performance has increased the pressure in many workplaces. UNISON believes that managers are increasingly bullying their employees in an attempt to drive them harder. It says that this is not only bad for the workforce but it is counter-productive as the most productive workplaces are those where workers are contented. In a recent survey (Rayner, 1997) by UNISON, two-thirds of members responding said they had experienced or witnessed bullying and employers were aware of three in four cases of repeated bullying. The survey also showed that 83 per cent of the bullies were line managers.

A 1995 survey of workplace reps by the Manufacturing Science and Finance Union (MSF, 1993) found that nearly a third of the representatives thought that bullying was a significant problem in their workplace yet, despite the size of the problem, three-quarters of employers had no policy for dealing with bullying.

TUC survey

In September 1998 the TUC's national opinion poll (NOP) survey was undertaken (TUC, 1998). This UK survey involved 1,002 people of working age across a wide geographic and social spectrum. The NOP survey asked employees whether

Table 13.3 TUC/NOP poll results: percentage of respondents who said they were currently bullied or had been bullied at work (extrapolated to represent the total working population)

Region	Proportion of working population	Region	Proportion of working population
Scotland	360,000 (9%)	North East	240,000 (12%)
North West	594,000 (11%)	Yorkshire and Humberside	320,000 (8%)
Midlands	1,036,000 (14%)	Wales	414,000 (18%)
East Anglia	137,000 (8%)	Greater London	495,000 (9%)
South East	778,000 (9%)	South West	390,000 (10%)

Table 13.4 TUC/NOP poll results: percentage of respondents who said they were aware of workplace bullying in a current or former job (extrapolated to represent the total working population)

Region	Proportion of working population	Region	Proportion of working population
Scotland	720,000 (18%)	North East	340,000 (17%)
North West	1,566,000 (29%)	Yorkshire and Humberside	1,240,000 (31%)
Midlands	1,924,000 (26%)	Wales	621,000 (27%)
East Anglia	447,000 (27%)	Greater London	1,485,000 (27%)
South East	2,333,000 (27%)	South West	780,000 (20%)

they had been bullied (see Table 13.3) and if they were aware of bullying in a current or former place of work (see Table 13.4). The survey showed that one in ten (11 per cent) of respondents were either currently bullied or had been bullied in the past. Over a quarter (27 per cent) were aware that colleagues had been bullied in their current or previous job. The survey also revealed that people in managerial and professional roles were more likely to have been bullied: 15 per cent compared with 10 per cent of skilled white-collar workers, 11 per cent of skilled blue-collar workers and 9 per cent of unskilled workers. According to the survey, men and women are equally likely to be the victims of bullying (11 per cent of men and 11 per cent of women).

Bullying is a health and safety issue

The effect of being bullied is frequently stress and ill health. Health and Safety Executive (HSE) guidelines (1998) identify bullying as a cause of stress and indicate that bullying must be taken into account when organisations undertake risk assessments.

These guidlines (HS (G)116) state that:

- employees cannot easily cope with inconsistency, indifference or bullying;
- employers must ensure that people are treated fairly and that bullying and harassment of those who seem not to 'fit in' is not allowed;
- employers should have effective systems for dealing with interpersonal conflict, bullying and racial or sexual harassment, including an agreed grievance procedure and a proper investigation of complaints.

Bullying at work must be dealt with as a health and safety at work issue. It cannot be regarded as an industrial relations issue, nor should it be left to the individuals concerned to settle. Employers have a clear duty under the Health and Safety at Work Act (1974) to ensure the health, safety and welfare of their employees. This duty of care is part of the contract of employment.

Employers will be regarded as failing in their duty of care if they do not anticipate the possibility of bullying or have procedures to deal with and resolve bullying once it has occurred. Employers must take steps to ensure that it is not their culture, management style or employment practices that enables individual bullying or corporate bullying practices to exist.

Bullying is clearly a major stressor in the workplace and again the Health and Safety Executive has made it very plain that stress should be dealt with under the Health and Safety at Work Act (1974) and under the Management of Health and Safety at Work Regulations (1999). Employers should carry out risk assessments to identify stressors within the workplace and trade union safety representatives have a right to be part of that assessment process. Attempting to carry out a risk assessment on bullying without a clear approach to a workplace stress policy would be impossible.

Workplace stress policies

An effective stress policy should:

- recognise that stress (and therefore bullying) is a health and safety issue;
- recognise that stress is about the organisation of work;
- be jointly developed and agreed with unions;
- have commitment from the very top;
- guarantee a 'blame-free' approach;
- apply to everyone;
- have arrangements for joint monitoring and the reviewing of effectiveness.

The three prime objectives of a stress policy are to:

- prevent stress by identifying the causes of workplace stress and eliminating them;

- recognise and deal with stress-related problems as they arise by educating employees about stress and encouraging participation and co-operative working;
- rehabilitate employees suffering from stress through the provision of independent confidential counselling and support.

What should unions look for in a bullying at work policy?

If the policy is to meet the needs of all employees the policy should include:

- a timetabled complaints procedure;
- training for management and trade union reps;
- recognition of the problems faced by the victim;
- provision of confidential counselling and support for the bullied and the bully;
- taking care of the supporters;
- monitoring and evaluation.

A timetabled complaints procedure

It is important that all those concerned know the nature of the procedure and at what stage the various elements of the procedure will come into play. For example, when the informal process has failed or ceased to be acceptable, at what stage will the disciplinary element of the hearing be invoked? The timescale must be as short as possible to ensure that the issue is quickly resolved. However the timetable must also ensure that all concerned are able to present their case fully. If there is to be a change to the timetable everyone concerned must agree to it.

Training for management and trade union reps

The main reason that bullies survive is that colleagues do not know how to confront the bully or to assist the bullied. Many people do not realise that bullying is occurring. A lot of bullying happens behind closed doors, without witnesses. Therefore the bully thrives where bullying is not talked about or where there is ignorance about the nature of bullying and how to deal with it. Bullying sometimes depends on the connivance of others not directly engaged in the bullying. Training programmes should not only deal with the implementation of the bullying procedure but also enable employees to understand the nature of bullying, the forms it can take and how they can recognise and assist colleagues with the bullying problem.

To be successful, training courses should be developed with the support of union or staff representatives. It is vital that staff are fully involved in the training programme and are not just seen as receivers of 'wisdom'.

Recognition of the problems faced by the victim

Employees experiencing bullying may not come forward as readily as employees concerned about other health and safety issues in the workplace. These employees may feel isolated and, if the bully is particularly manipulative, they may not even be certain that they are actually being bullied. Any policy on bullying needs to cater for those who do not feel confident about coming forward. It is important therefore to offer informal procedures to help victims of bullies.

The primary concern for anyone being bullied is for the bullying to stop as quickly as possible. Trade union representatives should be careful to check with the victim what they want to happen and only take a case further if the victim supports this action. Sometimes a quiet word with the bully indicating that there has been a complaint about their behaviour is enough to stop the bullying occurring.

Provision of confidential counselling and support for the bullied and the bully

Like all good health and safety procedures, the workplace bullying policy must prevent a reoccurrence of the event. Once the employer, in discussions with the safety representative and the workforce, has removed the organisational problems that led to the bullying, there may be a need for counselling. The bullied employees should receive counselling to help them cope with the aftermath of the experience of being bullied and also with having to go through the complaints procedure. Counselling can be used to rebuild their self-esteem. For the bully counselling can be helpful in enabling a recognition of the nature of their attitude or behavioural problem which led to the unacceptable behaviour towards others. The families of the victim and the bully may also need counselling to help them to come to terms with their experience.

Employee assistance programmes (EAPs), welfare services and other employee support programmes are often funded by employers to provide free and confidential personal counselling for employees and sometimes for their families. These services may be run in-house or externally. It should be remembered, however, that these programmes do not remove the problem and should never be regarded as a substitute for a preventative policy. Counselling is designed to help the employee deal with the aftermath of bullying. The Advisory, Conciliation and Arbitration Service (ACAS), which can be used in support of counselling, emphasises that its service should only be regarded as part of an employer's response.

Where counselling is requested or offered, it must be made available very quickly. No one involved in the bullying should have to wait more than a few days, regardless of whether he or she is the bullied or the bully.

Internal and external organisational counselling and support programmes offer a wide range of services. These often include:

- a free telephone service;
- assessments;
- face-to-face counselling;
- referral to expert advisers.

The decision to introduce a counselling programme and who is to provide that service are important questions and should only be made following discussion with the employees and their representatives.

Taking care of the supporters

It should be remembered that taking someone through a bullying complaints procedure can be a traumatic experience, not only for the bully and the person bullied but also for those providing the counselling, information and support. It is important to recognise that those who are personally concerned in the process, such as the trade union representative, personnel managers and counsellors, may become affected by dealing with particularly harrowing cases. The issues involved in providing this support are discussed in Chapter 11.

Monitoring and evaluation

Any policy and counselling referral must be subject to regular monitoring – is it achieving its intended aims and objectives? Just because there have not been any complaints since the policy or procedures were introduced does not mean that it is working well. It is important that the standards of performance and success criteria are agreed before the service is introduced so that everyone is very clear about what is required and the standards that can be expected.

Examples of good practice

Some of the TUC affiliates are already tackling workplace bullying and in the best initiatives the unions work with the co-operation of personnel or human resources. The Graphical Paper and Media Union (GPMU) has negotiated a 'Dignity at work' clause for members of the British Printing Industries Federation. The clause states that members must not be subjected to racial or sexual harassment or bullying.

The GPMU policy offers advice and guidance to officials in handling bullying or harassment. It makes clear that, where the guidelines are negotiated into agreements, they must cover all employees. This policy designates bullying behaviours as a disciplinary offence. While a grievance procedure is still part of the process, the policy also provides for an initial informal meeting where counselling is recommended.

In addition, the GPMU appoints lay women members within branches as liaison officers. These liaison officers act as a friend and point of contact for those suffering from harassment or bullying and provide a means of contact between the member and branch official. The liaison officers also assist branch officers in resolving harassment and bullying problems and can refer members to the counselling services when appropriate. The union believes that it was important for the liaison officers to be women in order to encourage women members suffering harassment and bullying to come forward without the additional ordeal of having to confide in a male official.

UNISON, the public sector union, became actively involved in dealing with bullying at work in 1980 following some serious allegations of bullying against a senior manager at Aberdeen City Council. These experiences in handling bullying helped UNISON to develop their first anti-bullying and harassment policy, together with a culture statement on dignity at work. UNISON commissioned a survey of their union representatives which revealed that 66 per cent had witnessed or experienced bullying and that 94 per cent said they felt bullies got away with bullying. Following the survey, UNISON produced guidelines for branches on tackling bullying and offered training to all their union reps.

The Transport and General Workers Union (TGWU) responded to some notorious bullying in the hotel and catering sector by developing an anti-bullying code of conduct which the Restaurateurs Association of Great Britain agreed to recommend to their members. The campaign was launched after the TGWU had worked closely with ITV's *The Big Story* that graphically exposed the bullying behaviours of some of London's top chefs.

The TGWU has put the responsibility for bullies firmly at the door of the employers:

> Employers often turn a blind eye to the actions of supervisors and department heads allowing them to create a climate of fear which means many workers will not complain – particularly as most of them have not worked long enough to claim unfair dismissal.

A special organising support unit has been set up by the TGWU to help employees stand up to bullying. The weekly sessions give employees – both members and non-members – the opportunity to get advice from the union about how to tackle workplace bullying. The TGWU found that the bullying had made many people too frightened to join a union. The advice sessions give employees the confidence to join the union that will be there to support them in their stand against bullying. The benefit for the employee is that the union organisation takes the focus off the individual and as a result the TGWU has been successful in getting bullying supervisors and managers to change their behaviour in a number of union-organised hotels.

What the role of a trade union representative should be

So far this chapter has dealt with what, from a trade union viewpoint, an employer's policy should look like and how it might operate. We now turn to look at the role of the trade union safety representative within that policy.

Safety representatives should *encourage employees to record episodes of bullying in writing*. It is important to establish an accurate record of what happened or is happening and the chronological order. Often the employee is too upset to have thought of making records of bullying behaviours.

Safety representatives should *find out if other employees have had similar experiences*. Many members who approach their reps believe that they are the only ones experiencing bullying. This is often not the case. There are serial bullies as the following case of bullying, brought to the attention of the National Association of Schoolmasters and Union of Women Teachers (NASUWT) by two members of staff in a LEA primary school, shows:

> She (the Head) would wait at the door in the morning and follow her victim around the school during breaks and at lunchtime, questioning them. It was a long time before the victims, both male and female, realised that the situation was one of bullying. Many staff were losing their abilities and several wept on many occasions. There seemed no one to turn to. When the union rep was approached, he turned out to be a victim also.

Next steps

Safety representatives should *only move as fast and as far as the member wishes to go*. The representative may realise she or he has a good case but the member may be in a fragile state of mind and may need a great deal of support. Even if the employee is determined, it should be made clear what is involved in going through the complaints procedure where their honesty and mental health may be challenged. The most important thing for the victim is to get the matter resolved as quickly and as quietly as possible.

Safety representatives should, *with the agreement of the member, report the incident to management*. Reporting the incident to management does not necessarily mean that a full-scale complaints procedure will be instigated. It does, however, ensure that management is made aware that there is a problem. Informal approaches should always be attempted to resolve the issue in the first instance.

Safety representatives should *represent the member at all stages of the enquiry*. One of the tools that the bully frequently uses is inconsistency. It is important therefore that, when dealing with a victim of bullying, the union is consistent in its approach. The same safety representative should accompany the member at all stages of the complaints procedure to ensure that the member knows and trusts the representative dealing with his or her case. A single contact also helps maintain confidentiality. Like any other union issue the member should never go to a meeting with management to discuss the case, however informal, without being accompanied by their representative.

Safety representatives should *ensure the case is dealt with quickly and in accordance with agreed procedures*. As already indicated, speed is of the essence in bullying cases and the complaints procedure should be enacted within days of the report of bullying being made. Any changes to the procedure or timing must be made with the agreement of the victim and the representative. The representative must make the implications of any such change clear to the victim and ensure that a return to the established procedure occurs as soon as possible.

Safety representatives should *contact the union if help is needed*. Bullying cases can be difficult and emotionally draining. A representative is not on their own and they are not expected to have all the answers. Contacting the union branch or regional office for advice and assistance with a bullying case is a sign of strength and not a sign of weakness. The union is there to help, support and advise representatives as well as ordinary members.

Raising awareness

It is not enough to have anti-bullying policies tucked away in files. Trade union representatives and personnel managers should get together regularly to introduce awareness-raising programmes for the workplace.

Workplace awareness raising could include:

- a survey of the workplace or branch on the extent of bullying;
- organising meetings on the issue;
- raising awareness through posters and leaflets;
- ensuring that bullying is incorporated into union and employer's training.

These are also activities that the union or staff association can carry out on its own behalf.

In 1996 Tesco introduced a 'Dignity at work' policy into its retail and distribution outlets. The policy was negotiated with the Union of Shop, Distributive and Allied Workers (USDAW) and was followed by training in all depots covering 7,800 staff (USDAW/Tesco, 1996). The company trained its managers and supervisors in the 'Dignity at work' policy and the union trained its representatives. This training involved all representatives being released for a two-day intensive training course. Depot employees were then required to undergo a training programme jointly conducted by USDAW and Tesco management.

Education and training on workplace bullying

Education and training is very important in developing the confidence and skills of safety reps in dealing with workplace bullying effectively. The TUC Education Service offers two- and three-day courses that look at workplace bullying from a trade union perspective so that union representatives can:

- define and recognise bullying, identify potential work organisation and staffing issues that may encourage bullying;
- work with personnel officers to ensure that bullying is addressed and managed;
- provide members with support where bullying occurs;
- understand how employment law principles and health and safety legislation apply to workplace bullying;
- develop a workplace strategy to tackle bullying.

The course offers credits at level 2 and 3 through the National Open College network.

Many unions also provide their own training for representatives and look at innovative ways of providing this training. BECTU (Broadcasting, Entertainment, Cinematograph and Theatre Union) has just become involved in an innovative approach to training their representatives in dealing with workplace bullying. This approach involved commissioning EQUITY (the actors' union) to run a role-play workshop on bullying. The EQUITY members played a bullying line manager and a new employee.

In a series of scenes the actors showed the steadily deteriorating relationship between the line manager and the new employee. Between each scene the characters were able to talk to the whole group, answer questions and explain their actions and feelings. The group suggested possible solutions – and then role-played these solutions.

Jill Lamede, course organiser, said:

> The members became totally involved with the emotions of these characters. Together we were able to experiment, to try out different approaches to the problem and see what worked or failed and why. I can't claim we had found a perfect solution to dealing with bullying but we certainly developed a greater understanding.

In addition to the specific courses aimed at bullying in the workplace, the trade union movement provides training to over 14,000 safety representatives each year. There are two blocks of training, each being one day a week over ten weeks: 7,000 go through the TUC schemes (TUC Education). This training provides safety reps with the knowledge to enable them to carry out their duties and to respond to the concerns of their members. Part of each course deals with stress in the workplace and the causes of stress such as bullying.

The TUC, in conjunction with the Industrial Society, has produced a video-based training pack *No Excuse: Beat Bullying at Work* (Ishmael, 1999). The video is set within the private and public sectors and vividly illustrates the different types of bullying behaviour, the effects and the outcomes. The aim of the pack is to provide organisations with practical advice and support which will assist them in combating bullying in the workplace. The basis of the training pack is that employers, staff and, where present, trade unions should work together to develop and put into practice the company policy on bullying at work. It is vital that employers recognise that training is a key element in their health and safety at work policy and in dealing with bullying at work. Union representatives must be allowed to exercise their legal right to time off, with pay, to attend these union safety courses. There are clear advantages to facilitating this training, as it will enable the rep to become an effective tool in assisting not only union members but also management in dealing with bullying and health and safety issues in general.

Unions and the law

The legal aspects of bullying in the workplace are covered in Chapter 7 of this book, however the unions approach to bullying is one of prevention rather than prosecution but, where appropriate, unions will seek damages from employers for injury, whether for physical or psychological, to their members.

Health problems and injuries sustained at work continue to make up the bulk of the union's legal work. Over the last five years, unions have won £1.5 billion in damages for its members who have been injured or made ill through work. In 1997, unions helped members to gain £28.5 million in compensation for occupational diseases. Employees who experience disease or accidents at work can apply for a variety of state compensation schemes or they can sue their employer through the civil justice system. Unions help members through both processes, providing legal expertise and covering the costs.

Unions approach stress and bullying the same as any other health and safety issue, as do courts and employment tribunals, and will take a member's case for bullying or stress to court if there has clearly been a breach of health and safety or employment law.

Going to court

When employees take their employers to court, they can sue for special damages: expenses arising from the accident or illness (for example, special care or adaptations to their home) and loss of earnings. Employees can also sue for general damages, that is, pain and suffering. The employee's lawyers must prove both that the accident or illness was caused by their work and, in the workplace (causation), that their employer was negligent in allowing the conditions in which the illness or accident happened (liability).

It is not enough for a person to feel that they are stressed or being bullied. Under the law, as it stands at the moment, the employer can only be held responsible if there is a clear detriment to a person's physical or mental health. This was demonstrated in the case of Walker vs Northumberland County Council (1995) taken by UNISON on behalf of their member, John Walker. Mr Walker suffered a nervous breakdown as a result of pressure at work. On his return to work the promised support did not materialise and he had a further nervous breakdown six months later. His employers were sued in the High Court for breach of duty of care. They were held liable for the second breakdown but not the first as this could not have been reasonably foreseen. The possibility of a second nervous breakdown was reasonably foreseeable. The case established that there is no reason why mental injuries suffered as a result of bullying or any other form of harassment should not be actionable. Mr Walker was awarded £175,000.

Another UNISON member, Janet Ballantyne, accepted an out of court settlement of £66,000 for the stress she suffered as a result of bullying at work. She claimed she was humiliated and countermanded by her supervisor in front of residents at the home for the elderly at which she worked, and that her employers did nothing to address the problem. Frequent criticisms led to Mrs Ballantyne taking days off work on grounds of ill health because she experienced panic attacks, for which she received medication. Eventually she suffered a major panic attack at work and was forced to take early retirement. She brought a claim against her employer for personal injury (Ballantyne vs Strathclyde Council, 1996).

To prove causation the union's legal team will often draw upon the knowledge of an expert witness such as a medical or psychiatric expert who will testify as to the nature of the illness and how it arose. In terms of liability, the lawyers have to prove that the company knew or should have known of the risks and that it could have taken reasonable steps to avoid them.

For the last four years stress cases have topped the list of cases taken to law by the unions. With the growing awareness of the relationship between bullying and stress-related illnesses and the large number of publications on this issue, including some from the Health and Safety Executive, employers cannot make the excuse that they were unaware of their duties to prevent bullying at work.

In a case where a widow of a UNISON member, who was driven to suicide by stress at work, received £25,000 in an out-of-court settlement from North East Essex Mental Health NHS Trust. UNISON described what had happened to the man as that he had been subjected to vindictive, oppressive, ruthless and macho-style management. His managers had been aware that he had been suicidal but had done nothing about it. This award made legal history as it was the first time a widow had received a settlement for suicide caused by stress at work.

The legal aspects of bullying in the workplace are covered in Chapter 7 of this book, however the union's approach to bullying is one of prevention rather than prosecution but, where appropriate, unions will seek damages from employers for injury, whether physical or psychological, to their members.

It is highly unlikely that a union representative will be required to deal with a legal case on their own beyond the company's internal procedures. The union's legal department would normally deal with these more complex cases. The union representative, however, will be an important source of information and, of course, a witness.

It is important that the representative supports his or her member throughout the case. This will involve acting as a consistent base, helping the member to cope with the great many stresses and strains involved in going through the legal procedures, dealing with lawyers and reliving the bullying experiences that brought him or her to this position.

There is not a clear, single piece of legislation dealing with bullying at work. However, it is worth emphasising that it is no defence for the employer to suggest that junior management or supervisors, to whom the task of providing a safe working environment had been delegated, were carrying out the acts of bullying. Under the Health and Safety at Work Act (1974) the task may be delegated, *but the employer's duty of care is not.*

A case of bullying can be brought through either the employment tribunal (ET) or the courts. There are advantages in going through the courts as a case can be lodged in court up to six years after the event. This compares with only three months for ETs. Employees who bring cases to court may also be eligible for the payment of legal aid, which is not the case in ETs. However, legal aid is not an issue for a union member, where their union takes up the case.

There are a number of advantages to going through the ET procedure. The panel is likely to be more aware of the industrial/workplace context as it is made up of a lawyer and representatives from the trade union and employers. It is also possible in ETs to include other issues related to the complaint, for example sexual and racial harassment or other matters which led to the person leaving their job. This flexibility is not possible in the court system where complaints are normally confined to the actual complaint that has been lodged.

A final word on going to law

Many people who have been bullied at work want to make sure that what they experienced never again happens to a colleague or new employee. This is why most people are prepared to go through the grievance procedure or tribunal and court procedures. Very few victims of bullying seek revenge. Standing up to bullying can be a long and painful process and the success can never be guaranteed. It takes a very strong person to withstand the process and to overcome the disappointment should the result not be what they had expected. There are times when the union will advise a victim of bullying that, despite all the pain and suffering that has been endured, the case is not strong enough in law, for even a 50/50 chance of success. In these cases the union advice may be to drop the case. This advice should always be considered carefully by the victim. There comes a point when it is more important for the employee to look after their health and well-being, and that of their family and friends, and to recognise that there is nothing more to be gained from further action. In these situations all that can be done is to come to terms with what has happened and to move on. This can be one of the most difficult decisions a victim of bullying has to make. The problem that causes this painful dilemma is the lack of clear legislation that deals with bullying at work. The TUC and its affiliated unions supported the 'Dignity at work' legislation that was proposed by Lord Monkswell prior to the 1997 General Election. It is the view of the TUC that the law in this area is still in need of serious review.

Conclusion: a final checklist on bullying

Policies should include:

- a statement that bullying will not be tolerated and will be treated as a disciplinary offence;
- a commitment that complaints of bullying will be taken seriously and dealt with quickly and in confidence;
- a complaints procedure with a timetable for all stages of the procedure;
- a training for management and trade union reps;
- a provision of confidential counselling for both the bully and the bullied;
- regular monitoring.

Safety representatives should:

- encourage members to record episodes of bullying in writing;
- establish whether other employees have had similar experiences;
- discuss with the member the various options and always be clear on the steps both agree to take next;
- report the incident to management, with the agreement of the member;
- represent the member at all stages of the enquiry and ensure that the case is dealt with quickly and in accordance with the agreed procedures;
- contact the union, if additional help or support is needed.

Campaigning could include:

- a survey of the workplace or union branch on the extent of bullying;
- organising meetings on the issue;
- raising awareness through posters and leaflets;
- ensuring that dealing with bullying is incorporated into union and employers' training programmes.

Acknowledgements

I would like to take this opportunity to thank the many people who, through their support for the TUC's campaign 'No excuses: beat bullying at work', made my contribution to this book possible. In particular I would like to thank Tom Jones and Thompsons Solicitors, who sponsored the TUC campaign and who have been of considerable support, also my colleagues at the TUC: Steph Power who was the drive behind the project and Liz Rees who helped develop the training pack. Our thanks also go to the Industrial Society for a successful collaboration.

Thanks are also due to Lyn Witheridge and her colleagues at the Andrea Adams Trust, with whom the TUC is pleased to be associated, and to Dr Cary Cooper, Professor of Organisational Psychology at UMIST, for his support of the unions in their work on workplace stress and bullying. Thanks also to the many trade union safety specialists and trade union representatives who have not only contributed to the TUC's work on bullying but also support their members on a daily basis in difficult circumstances. And last but not least to Helge Hoel, Cary's colleague at UMIST, and other colleagues on the British Occupational Health Research Foundation's 'Bullying at work' project steering committee, whose ideas I have pilfered.

References

Adams, A. (1992) *Bullying at Work: How to Confront and Overcome It*, London, Virago.

British Printing Industries Federation, internal document, London.

Graphical Paper and Media Union (GPMU), internal document, Bedford.

Guardian, (1 December 1998) 'Managers under stress: new workplace bullies', p. 7.

Health and Safety at Work Act (1974), section 2 (1) London: The Stationery Office.

Health and Safety Executive (1998) *Help on Work-related Stress: A Short Guide*, INDG281, Sudbury, Suffolk: HSE Books.

Jenkins, R. and Warman, D. (1993) *Promoting Mental Health Policies in the Workplace*, London: HMSO.

Jones, J. R., Hodgson, J. T., Clegg, T. A. and Elliot, R. C. (1998) *Self-reported Work-related Illness in 1995: Results from a Household Survey*, London: HSE Books.

Kirby, P. (1998) 'Twenty-one years of saving lives', TUC survey of safety representatives, London: TUC Organisation and Service Department.

Institute of Management (1996) *Are Managers Under Stress?: A Survey of Management Morale*, London: The Representation Unit, Institute of Management.

Ishmael, A. (1999) *No Excuse: Beat Bullying at Work*, a video-based pack on how to tackle bullying in the workplace, London: The Industrial Society.

Management of Health and Safety at Work Regulations (1999), London: The Stationery Office.

MSF (1993) *Bullying at Work: How to Tackle It*, a guide for MSF representatives and members, London: MSF Centre.

Rayner, C. (1997) *Unacceptable Behaviour: Workplace Bullying Survey*, Woolwich: UNISON Health and Safety Unit.

TUC Education, TUC, Great Russell Street, London.

TUC (1998) 'Five million bullied at work', press release at start of 'No excuses: beat bullying at work' campaign, 5 October 1988, London: TUC Campaigns and Communication Department.

Tudor, O. (1996) *Stressed to Breaking Point: TUC Survey of Safety Representatives*, London: TUC Organisation and Service Department.

UNISON (1997a) *Bullying at Work: Guidelines for UNISON Branches, Stewards and Safety Reps*, CD/27/403, Woolwich: UNISON.

UNISON (1997b) 'Bullying at work', a pamphlet for representatives and members, Woolwich: UNISON.

USDAW/Tesco (1996) 'Dignity at work agreement', Manchester: USDAW; Waltham Cross: Tesco Stores Ltd.

Walker vs Northumberland County Council (1995) IRLR 35, QBD.

CHAPTER FOURTEEN

Building a culture of respect

Issues for future consideration

NOREEN TEHRANI

Introduction

This book has been concerned with looking at the way that organisations can foster a culture in which individual employees respect each other and in which the position of the organisation is both respected and respectful. Unfortunately this ideal is not always achieved and where relationships break down it is essential that the organisation has mechanisms to deal with the perpetrator and the recipient of bullying. Owing to the complexity of human relationships organisations are frequently faced with circumstances in which there is clear evidence of interpersonal conflict, and it is difficult to be certain who is the bully and who is the bullied.

Bullying defined

Defining bullying is not a simple task. With the large number of definitions of bullying that have been created, it is not surprising that each worker or researcher in the field chooses a definition that most closely reflects their interests or theoretical orientation. Hoel and Cooper use a definition of bullying that brings together many of the accepted features of bullying. The definition requires the bullying behaviour to be of a negative nature, persistent, long-term and involving an imbalance of power. Even if it were possible to agree on the nature of bullying there is a wide range of views on why bullying occurs. The view held by Crawford is that everyone has the capacity to become a bully but the expression of bullying behaviour is strongly determined by the nature of the working environment. The idea that the target of bullying being a passive target of negative behaviours is rejected by Hoel, Cooper and Crawford who recognise that each negative act is likely to produce a response from the recipient. This response can, in some circumstances, begin a process of escalation in which the behaviours become more and more extreme. While an agreed definition of bullying may appear to be essential, it is important that the definition recognises the complexity of the interactions between the perpetrator, target and the organisation.

The impact of bullying

General and occupational health professionals are often the first professionals to become aware of workplace bullying. Manning looks at the role of the GP in identifying where a patient's symptoms are caused by interpersonal conflict and describes how the GP can support the patient in dealing with the medical and psychological outcomes of bullying (Chapter 12). While it is well recognised that victims of bullying frequently suffer mental and psychosomatic health consequences, there is a growing body of evidence to show that the symptoms experienced by some victims are consistent with a diagnosis of post-traumatic stress disorder (Leymann and Gustafsson, 1996).

Scott and Stradling use the diagnostic criteria used to assess post-traumatic stress disorder and argue that while it is not possible to make a formal diagnosis of this disorder, in many people the responses to exposure to bullying are indistinguishable from those caused by an exposure to a traumatic event. Victims of bullying have difficulty in understanding their experiences and frequently experience symptoms of re-experience, hyper-arousal and avoidance. These symptoms are graphically described in the personal narrative accounts presented by Tehrani in Chapter 4. While attention has been drawn to describe the causes and impact of bullying, much less has been done to develop effective interventions to help both the perpetrator and the target.

Organisational and social issues

An issue common to several contributions is the fact that we often fail to examine the impact of bullying within the wider organisational and societal context. Hoel and Cooper provide an overview of these issues in Chapter 1, these are further developed by Lawrence in Chapter 5 with regard to social interactionist theories, and how these theories can enrich our understanding of the way that expectations and perceptions of individuals involved in bullying can affect the outcome. In Chapter 9 I provide some tools to help organisations create an environment that would deal with interpersonal differences and conflict as ways to improve the quality of relationships, rather than brush the problems under the carpet. The description provided by Walker in Chapter 8 describes the way that a health trust in Northern Ireland proactively worked to build a culture which supported the dignity of its workforce. Walker provides a working example of how changes to the organisation can bring about changes in the behaviours of the organisation and its employees. Making changes in the way that organisations behave is never easy and must be actively supported by enlightened leaders. It is important, therefore, that efforts are made to quantify the benefits of creating a culture of respect in terms of productivity and performance, as well as in terms of employee satisfaction and attitudes.

Making the difference

If organisations want to make a difference, it is essential that they have some idea of the incidence of bullying. In a review of the findings of a large number of studies into bullying, Beale in Chapter 6 comes to the view that there are difficulties in comparing and contrasting the findings owing to the different ways in which the surveys were constructed. The use of employee surveys within organisations was found to be of more value by Rains (Chapter 10) who, in an attempt to quantify the size of the problem in the Post Office included questions on bullying in the employee opinion survey. Despite the introduction of equal opportunities' policies and training, Rains was dismayed to find that only 47 per cent of the workforce believed that bullying was taken seriously. Beale describes the need for organisations to monitor two aspects of workplace bullying as a way of making a difference. Firstly, the need to monitor the extent and nature of the bullying that is occurring and, secondly, to track the way individual cases of bullying are dealt with by the organisation. Monitoring the incidence of workplace bullying will never be easy, but it is important that organisations begin the process of identifying the nature and impact of bullying.

Formal procedures and the law

The use of grievance and disciplinary processes are used to investigate and determine whether bullying has occurred, as well as to decide on what actions need to be taken to punish the perpetrator and protect the victim. While the use of formal procedures must always remain an option, the use of informal procedures has been found to be more successful in achieving settlements which meet the needs of all parties. Bullying is not always recognised when it first begins and therefore by the time that it becomes obvious that a particular behaviour is bullying, the relationship between those involved is likely to have broken down to a point at which reconciliation is difficult, if not impossible, to attain. Mellish (Chapter 13) describes the approaches adopted by the Trades Union Congress and Leighton (Chapter 7) describes the current status of the law. Many employees and organisations that have resorted to the law to resolve interpersonal conflict have found that they cannot achieve their desired result even when the court's decision is in their favour. While the use of formal procedures to reinforce the seriousness of bullying in the workplace has an important role to play in dealing with the problem, it can never be the total answer to the problem. Mellish highlights the need for increasing manager skills in dealing with employee differences and conflict and the need for organisations to undertake constructive mediation and conciliation involving the individuals, union and the organisation as a matter of high priority.

Support and supporters

Dealing with bullying in the workplace can have a marked impact on everyone involved. Proctor and Tehrani use their chapter (Chapter 11) to describe how the dynamics of bullying can shift and the supporter is placed in a position that can cause them to become a victim or a bully in the case. Both Walker and Rains describe the importance of the support that their organisations have provided for 'peer supporters' and other 'helpers' within the organisation with a responsibility for dealing with the aftermath of bullying. It is essential that organisations that use 'peer counsellors' or 'peer supporters' ensure that these people have the appropriate training and support to undertake this difficult and demanding work.

Conclusions

Overall this book identifies that there is a need for a practical, conceptual and methodological shift in the way that bullying is viewed in the workplace. Instead of organisations and researchers spending all their time in defining and describing which behaviours can be categorised as bullying, the emphasis should be shifted towards defining and describing how organisational cultures can respect the rights and dignity of all employees. However, this change will require a conscious shift from the current emphasis on punishment to one in which the rewards for positive interpersonal relationships are rewarding and rewarded.

Reference

Leymann, H. and Gustafsson, A. (1996) 'Mobbing at work and the development of post-traumatic stress disorders', *European Journal of Work and Organizational Psychology*, 5, 2, 251–75.

Index

absenteeism 77, 87, 88, 100, 101, 190
abuse: emotional 77
Adams, Andrea 24, 83, 155, 201
Advisory, Conciliation and Arbitration Service (ACAS) 110, 113, 198, 209
advocate: GP as 193–4
affective aggression 69
aggression: affective 69; covert 81; definition of 61–2; distinctions between bullying and 61, 63–9; escalatory model of 66–8, 73; instrumental 69, 70; overt 81; role of 22; social interactionist theory of 61, 63, 65, 70, 71–2, 73, 222; theories of 6–7, 61–3, 82
alcohol: abuse 192
alcoholism 197
Alemoru, B. 79
alternative dispute resolution (ADR) 110, 113
Andersson, L. M. 79
Andrea Adams Trust 83
anger control 38
anxiety 186, 190–2, 197, 201; -depression 188
apathy: bystander 72–4
Aquino, K. 78
Archer, D. 11
Ashforth, B. 78
assault 106–7
attribution theory 8, 15
avoidance 38
awareness: raising, 213

balance model of stress 38–40
Bandura, A. 71–2
Barling, J. 81
Baron, R. A. 81, 82
Beale, Diane 89, 223
behaviours: classification of bullying, 80–1; expectations of 121
benchmarking 147–8, 149, 151, 156
beneficial triangle, the 176–7
best practice 147–8, 156
Bjorkqvist, K. 82, 83, 88
Boal, C. K. 34
boundaries 169
brainstorming 143–4
Brodsky, C. M. 69, 70
bullied as bully, the 27, 99
bullying: classification of bullying behaviours 80–1; expectations of, 121; definitions of 4–5, 62–3, 79, 98, 99–102, 128, 205, 221–2; levels of 6–15; as an umbrella term 16
burnout 178–9, 187, 193
bystander apathy 72–4

care: standard of 106
caring for the bully 196–7
case studies 25–6, 28–9, 35, 140–1, 142, 146, 171, 172, 173, 176, 187–9, 191–2, 194–6, 204
causes of bullying 3–16, 222
Chartered Institute of Personnel and Development 119, 166
choice 174
cognitive-behavioural treatment 37–40
Coie, J. 11
common law 101
communication 185; quality of 165, 171–2, 186–7
comparator approach in discrimination law 99
compassion fatigue 178–9, 187
compensation 105, 111, 124
complaint 127; formal 88, 156; procedure 208, 209, 210, 212–13, 218
conciliation 113, 129, 146–7, 167, 170, 224
conduct: standards of 121
confidentiality 112, 170, 213
consultative support *see* supervision

conflict escalation 9–10, 221–2
context 65, 129
contingency fee arrangements 109
continual improvement: approach 136–7, 153; tools for 143–5
contract: law of 101; of employment 102, 103, 108, 207
contracts: psychological 168; supporters 168, 169–71
Coomber, S. J. 120
Cooper, Cary L. 7, 65, 66, 82, 83, 85, 87, 205, 221, 222
coping strategies 40
Corey, G. F. 178
corporate bullying 203
cost benefit analysis 147
counter-transference 178
covert aggression 81
Crawford, Neil 24, 221
Criminal Justice and Public Order Act (1994) 30
cultural antecedents of bullying 11–13
culture of respect 126, 129–30, 132, 135–53, 168, 173–4, 181, 182, 221–4
cycle of violence 7

Dearlove, D. 120
debriefing 43–4
de-individuation 72–3
Deiner, E. 72
depression 35, 36, 38, 106, 186, 187, 190–3, 197
Depression Alliance 193, 198
desensitisation 38
diary: keeping a 195–6
dignity 123, 124–5, 129, 132, 137, 140, 153; at work 98, 112, 117, 118, 125–30, 203, 210, 211, 213, 218, 222, 224
Dignity at Work: Bill (1966) 29–30; (1999) 103; European Commission recommendation on, (1991) 98, 101–2, 103
discretion 98
discrimination: laws on 99, 101, 103–5
dispute-related bullying 9
diversity 123–5, 126, 127, 128–9, 137, 150
Dodge, K. A. 11
downward bullying 82
drama triangle, the 175–6
duration of bullying behaviour 81
dyadic level, examination of bullying at 3, 7, 8–10, 11
dynamic aspects of the bullying process 7, 16

education on workplace bullying 214–15
Einarsen, S. 4–5, 7, 9, 16, 77, 78, 79, 80
e-mail 27, 99
emotional abuse 77
employee resistance 14–15
employee surveys *see* staff surveys
employee welfare services 165
employment-related stressors 34, 39–40
Employment Rights Act (1996) 103
enforcement 98
equal opportunities 123–5, 137, 156, 223
Equal Opportunities Commission 126
escalatory model of aggression 66–8, 73, 221–2
European Foundation of Quality Management, the (EFQM) 136, 146, 151–2, 153
evidence 110; lack of 157
exclusion: social 4
extent of bullying 85–90, 92, 205–6

Fair Employment Commission 126
fault 97
Felson, R. B. 9
Figley, Charles 178, 179
Fire Service: bullying in 11–12, 13
flame-mail 27
focus groups 141, 142, 143, 151, 157
Folger, R. 81
formal complaint 88, 156, 157
frequency of bullying behaviour 81
Freud, Sigmund 30

Geddes, D. 81
gender 82–3
general practitioner (GP) 185–98, 222; as advocate 193–4; effect of litigation on the 189–90; stress 186
generic harassment 78
Glover, Edward 30
government action contributing to bullying 15
graffiti boards 145

Graphical Paper and Media Union, the (GPMU) 210–11
Greenberg, L. 81
group level, bullying at 3, 10–11, 72–3, 99
Gustafsson, A. 43

harassment 4, 38, 40, 62, 77–8, 82, 91, 98, 99, 105, 107, 125, 127, 128, 129, 131, 141, 142, 155–63, 193, 196, 211; generic 78; racial 5, 78, 210, 217; sectarian 128; sexual 5, 78, 146, 210, 217; support officers 125–7, 130
harm 98, 100
Haskel, Lord 30
head-on approach 28–9
health and safety: advisers 92; approach to tackling bullying 77, 86, 91, 201, 206–7, 209; inspectors 98, 101, 108; legislation 98, 101, 103, 107, 108; policies 112
Health and Safety at Work Act (1974) 207
Health and Safety Executive (HSE) 206
Hellesøy, O. H. 79
Hoel, Helge 7, 62, 65, 66, 80, 82, 83, 85, 87, 221, 222
horizontal bullying 82

ill-health: bullying as a cause of 188–9, 191, 201, 205, 206; in care professionals 192–3
incivility: workplace 79
individual level, examination of bullying at 3, 6–8, 91–2, 203
individual repair 40
Industrial Society, the 79, 83, 89, 121, 122, 198, 215
instinct: to bully 22
Institute of Management 205
institutionalisation of bullying 11, 13
instrumental aggression 69, 70
instrumental character of bullying 14, 69–71, 73
intent: role of 5
intentions 83–5
investigators 26–7
Investors in People (IIP) 136, 152–3
Ishmael, A. 79

Jackson, S. E. 179
Johns, N. 13
Joyce, M. E. 148

Karpman, S. 175
Kay, John 131
Kotter, J. P. 132

Labour Relations Agency 126
Lancashire County Council 79
Larson, Ernie 197
law: civil 107; common 101; criminal 107; European Union 101; of contract 101; of negligence 101; spirit and letter of the 122–3; trade unions and the 215–18; traditions of British, 98
Lawrence, Claire 63, 64, 222
leadership 12, 152, 223; competencies 136–7
Leather, P. 63, 64
Lee, D. 14
Lees-Haley, P. R. 84
legal aid 109, 113
legal framework 98–102
legal processes 98, 109–11
legal system 97; workings of 102–8
Leighton, Patricia 223
Leymann, H. 7, 43, 81
liability 97
Liefooghe, A. P. D. 8
lies 24
listeners scheme 155–63
listening 186–7
litigation 97, 109–11, 112, 147, 215–18, 223; culture 97; effect of, on GP 189–90

Management of Health and Safety at Work Regulations (1999) 86, 107
managing by process 145–6
Manning, Chris 222
Manufacturing and Finance Union (MSF) 29, 205
maps: process, 145–6
Maslach, C. 179
Matthiesen, S. B. 9, 79
McCarty, Lord 29–30
McDonald, J. J. 84
mediation 88, 110, 113, 129, 146–7, 167, 224

medical consequences of bullying 185–98
medicalising problems 185, 197
Meichenbaum, D. 38
Mellish, Tom 223
Mentzel, P. J. 13
minority groups 10
mission statement 117–19, 120, 121, 129, 130, 136
mistreatment 77
mobbing 77–8
modelling 71
monitoring 136, 147–8, 208, 210, 218; benefits 141–2; bullying 77, 86–90, 92, 223
Monkswell, Lord 29, 218

narcissistic personality disorder 84
narrative therapy 43–58, 222
National Association of Schoolmasters and Union of Women Teachers (NASUWT) 212
negligence: law of 101, 103, 106, 109, 111
negotiating skills 169
Nesler, M. S. 67, 68
Neuman, J. H. 81
norms 65–9, 72
Novaco, R. W. 64
Novell UK 27

obsessive–compulsive disorder 84
occupational health practitioners 87, 92, 173, 222
Olafsson, R. 8
Olweus, D. 63, 70
open questions 186
organisation values 117, 119–23, 124, 130, 131–2
organisational change 13
organisational dysfunction: bullying as a symptom of 23–4
organisational expectations 168
organisational level, bullying at the 3, 11–13, 15, 65, 71, 81, 91–2, 222–3
organisational repair 40
overt aggression 81
Oxenburgh 147

paranoia 84
Pearson, C. M. 79
peer rejection 11
perceptions of bullying 83–5, 87
personal derogation 4
personal power 177–8
personality disorders 84
personality traits of those involved in bullying 3, 6–7, 9, 10, 15, 83
petty tyranny 78
policy 91, 127, 140, 145, 150, 152, 156, 157, 160, 167, 196, 208–10, 212, 218; development of 24–5, 112, 155, 201; workplace stress 207
political correctness 22
post-traumatic stress disorder (PTSD) 33–41, 43, 99, 222; symptoms of 33–4
power 176; abuse of 12, 78, 201; balance of 5, 68, 69–70, 73, 81, 82, 87, 221; personal 177–8
predatory bullying 9
preference stating 169–70
preventative action 86, 91, 107
probity 118; research 119–23, 129, 130
problem-solving 143, 195
procedures 87, 91, 112, 124, 127, 140, 158, 167, 209, 210; development of 24–5, 155, 201
process: managing by 145–6; maps 145
processes: informal 157; legal 98, 109–11
Proctor, Brigid 224
prolonged duress stress disorder (PDSD) 36–8
Protection from Harassment Act (1977), the 78, 103, 107
provocative victims 11
psychodynamic therapy 10, 178
psychopaths 6, 16
punishment 24, 68, 72, 124, 223, 224
purpose stating 169

quality improvement cycle, the 137, 140
Quine, L. 80, 82, 85, 190–1

Race Relations Act (1976) 104
Rains, Steve 223, 224
Raknes, B. I. 79
Randall, Peter 6–7, 27, 63, 68, 70, 71
Ravin, J. 34

Rayner, C. 7, 16, 62, 68, 73, 80, 82, 84, 85, 87
recognition: of achievement 148, 151; of problems faced by the victim 208–9
rehabilitation 91
reinforcement 71, 130
repair: individual 40; organisational 40
reporting bullying 86, 88–9, 125
Reporting of Injuries, Diseases and Dangerous Occurrence Regulations (1995) 88
resistance: employee 14–15
respect: culture of 126, 129–30, 132, 135–53, 168, 173–4, 181, 182, 221–4
restaurant kitchens: bullying in 12–13, 211
Richman, J. A. 82, 86, 88
rights at work 98, 101, 224
risk assessment 86, 87, 92, 107, 206, 207
road rage 30
role-conflict 12
roles: organisational 167; of supporter 170–1
Royal Mail 155–63, 223

scapegoating processes 10
Schultz, Will 174, 177
Schuster, B. 11
Scott, Michael J. 43, 222
self-assessment 148–51
self-help groups 193
self-regard 174
Sex Discrimination Act (1975) 104
Siann, G. 62
situational factors 7, 11–13
skills: supporter 171–2
Smye, M. 191
social desirability response bias 90
social exclusion 4
social incompetence 11
social interactionist theory of aggression 61, 63, 65, 70, 71–2, 73, 222
social learning theory 6–7, 71–2
social representations 8
social stress 11, 78
socialisation processes at work 11–12
societal level, examination of bullying at 3, 13–15
solution-focused interventions 195
staff: appraisals 90; surveys 90, 92, 141–2, 144, 157, 162, 223; turnover 77, 87, 91, 100, 190
stakeholders 135
standard: of care 106; setting 120
Stradling, Steven G. 43, 222
stress 188–9, 190–1, 193, 194, 197, 201–4, 205, 206, 214; balance model of 38–40; GPs 186; -mapping 195; social 11, 78; workplace, policies 207–8
stressors 191; employment-related 34, 39–40, 86, 207; social 78
suicide 197, 216
supervision (consultative support) 179–82; peer 180
Supervision Alliance Model 179
support: provision of 208–9; roles 170–1; sources of 165, 166–7, 224; for supporters 26
supporter skills 169–70, 171–2

surveys: staff 90, 92, 141–2, 144, 157, 162, 223
survivor's psalm, the 58
survivors 47, 50, 53, 57–8

team level bullying 91–2
team work 173
technological change 13
Tedeschi, J. T. 9, 67, 68
Tehrani, Noreen 222, 224
Thylefors, I. 10
total quality approach 135–53
trade unions 14–15, 28, 29, 85, 87, 109–10, 125, 126, 129, 141, 160, 167, 196, 201–19, 224; representatives 26, 87, 129, 159, 160–61, 173, 196, 201, 202, 207, 209, 212–15, 217, 218
Trades Union Congress 100, 189, 198, 201, 202, 203, 205–6, 210, 214, 215, 218, 223
traditions of British law 98
training 28, 208, 214–15, 218, 224
transactional theory 9
Transport and General Workers Union (TGWU) 211
trauma counselling 43
trespass to the person 101, 106–7, 109

tribunals: employment 108–9, 111, 112, 124, 131, 141, 147, 217
truth 24, 174
turnover: staff 77, 87, 91, 100, 190

UK National Workplace Bullying Advice Line 198
Union of Shop, Distributive and Allied Workers (USDAW) 213
University of Manchester Institute of Science and Technology (UMIST) 205
University of Ulster 125
upward bullying 82

values: organisation 117, 119–23, 124, 130, 131–2
Vartia, M. 12
victimisation 5, 77–8, 98, 103
violence: cycle of 7; work-related 78, 86, 91

Walker, Vivienne 222, 224
welfare services: employee 165
Welfare Workers Association 166
Welsh, W. N. 64
whistle-blowing 106, 193
Woolf reforms 110, 112
workplace incivility 79
Wright, L. 191

Zapf, D. 12, 83
Zillman, D. 64
Zimbardo, P. G. 72